Welcome
to the Farm

Welcome to the Farm

HOW-TO WISDOM *from* THE ELLIOTT HOMESTEAD

SHAYE ELLIOTT

GUILFORD, CONNECTICUT

An imprint of Globe Pequot

Distributed by NATIONAL BOOK NETWORK

Copyright © 2017 by Shaye Elliott

Photographs on pages i, ii, vi, x, xiv–xv, 114, 118, and 314 by Mary Collier.
Photograph on page 13 © iStock/rbiedermann
All other photographs by Shaye Elliott
Chapter opening illustrations by Farmrun
Spot art throughout © iStock.com/cat_arch_angel

British Library Cataloguing in Publication Information Available

Library of Congress Cataloging-in-Publication Data Available

ISBN 978-1-4930-2601-2
ISBN 978-1-4930-3042-2 (e-book)

∞™ The paper used in this publication meets the minimum requirements of American
National Standard for Information Sciences–Permanence of Paper for Printed Library
Materials, ANSI/NISO Z39.48-1992.

Printed in the United States of America

Dedicated to all my wonderful blog readers
who inspire me daily to keep writing,
creating, and farming.

CONTENTS

INTRODUCTION

Finding the Way to the Farm

"I think we should buy this cow," I proclaimed to my husband from behind the computer screen. Maybe if I didn't make eye contact, I thought, he wouldn't be able to glare at me for suggesting such a ridiculous proposition. After all, we'd just spent a year barely scraping by–cutting all but the essentials out of our budget and living more frugally than we ever had before. If the budget did give us any extra wiggle room, which it never did, it certainly wasn't going to be spent on a dairy cow.

Oh, by the way, did I mention that we didn't have a farm? At the time, we lived in a small old fishing house that was nestled in a quiet neighborhood in Southern Alabama. Our entire front yard was filled with sand from Mobile Bay, there wasn't a patch of grass to be found, and our nearest neighbors were about thirty feet from my window. I'm sure they would've loved waking up to a bellowing dairy cow. What exactly was my grand plan, anyway? To walk the cow down South Winding Brook Drive and tie it up on our front lawn? Let's not focus on why I was even browsing dairy cows on Craigslist in the first place, because that's not the point. The point is that I was sitting here, eyes wide with wonder, dreaming of what it would be like to be the owner of such a magnificent beast. To milk your very own cow! Can you even imagine? I could. That's all I could do–spend far too much time dreaming of the farm life that had taken up residence in my heart, unyielding to practicalities and wisdom. Hence the not-thinking-before-speaking situation in which I had just found myself.

"Honey," he replied softly (he's so sweet), "we don't have a farm." I knew this. "And we don't have the money." I knew he was right (don't tell him I said this, but he always is). Unfortunately, for him, this bud in my soul was beginning to blossom. It was coming to life with every afternoon

spent in the garden, every home-cooked meal, every glass of raw milk, carton of local eggs, and cow on Craigslist. I can't pinpoint the exact moment I decided I wanted to be a farmer. All I knew was that it was happening. Soon, I was hanging laundry over the fence line to dry in the summer sun, adding meat rabbits to the backyard, and planting kale in pots on the front porch, and I began to work toward a farm that, at the time, only existed in my mind.

I could feel it . . . taste it . . . smell it. It was there, ready for me to bring it into fruition. Do you feel the same? Is there something about a flannel shirt and basketful of tomatoes that makes you feel at home? Something tugging at your soul a bit, reminding you that there's a piece of the world out there that's real and raw and glorious?

As the good Lord would have it, we did buy that cow. Right before moving two thousand miles across the country to our first farm, in the Pacific Northwest, to welcome her home. She arrived at our barn before

we had even a single fence post in the ground, and the fact that my family didn't completely disown me for putting them in that situation still amazes me. I made many mistakes in my first few years of learning how to "farmgirl," not the least of which was that cow. Yet no matter how many mistakes I made, and no matter how many tough situations I had to work through on the farm (and there were many), I still woke up with a fire in my
belly to keep striving. I was hell-bent on chasing that beautiful rainbow of a life that promised a connection not only to the earth, but also to a community of people who slaved in the soil and experienced the ebb and flow of life with livestock. Because to this farmer, it mattered.

It mattered to me how my meat was raised. It mattered to me how it died, how it was treated, and how the product was managed. After years spent in commercial meat production, it became a huge relief to me that I could be in charge of exactly how my meat was raised and killed.

It mattered to me how my food tasted. It mattered to me how it was grown, harvested, and preserved. It mattered to me where it was grown, how long it took to get to my plate, and if the farmer received a fair price for his labor of love. It mattered to me that I was connected to the very lifeblood that sustained me day in and day out.

And it mattered to me that my kitchen table displayed a fresh bouquet of zinnias from the garden.

When I came to the farm, I came home.

The History

Neither my husband nor I grew up on a farm. I remember plucking green onions from my grandpa's garden that he meticulously kept alongside his old brick house, but that's about as hardcore as it was in those days. This same grandpa also welcomed me as he canned homemade applesauce and

pears each fall from the orchard that sat behind his house. It wasn't until my late teen years that I was introduced to the idea of real "farming" when my boyfriend allowed me to tag along as he trained and prepared a steer to raise and sell at the county fair. Driving down country roads in a pickup truck all of a sudden made more sense when I was wearing a sundress and cowgirl boots. We were, after all, going to shovel out manure from the barn. How hopelessly romantic!

Benny was that steer's name. The steer that made me fall in love with bovines. In the years that followed, I began to raise my own animals for the county fair and got a taste of the country life that seemed to be calling for me with its hay bales and from-scratch biscuits. It spoke so true to me that when it came time for college, I majored in Animal Science, Beef Production. My hope was to marry a cowboy and move to a big ol' fancy ranch, where we would raise a gigantic herd of cows and ride off into the sunset and all that jazz. Still hopelessly romantic, right?

As the good Lord would have it, I didn't end up marrying the cowboy. Rather, I met a Southern man at a bar who was simply passing through town on his way to something familiar, planned, and known. We fell desperately in love that night (who could've resisted my pink bedazzled "Bridesmaid" tank top?) and though he didn't share my affinity for farm life at the time, he was eager to listen to my wild dreams.

He introduced me to Pink Floyd and I introduced him to horseback riding. We shared evening rides through the local orchard to grab apples from the trees. We began to talk about vineyards and homemade wine. He watched me learn to cook from scratch and ate all my disastrous meals with a smile. He was the man I wanted to build my farm with. And so we did. Almost a decade and four children later, we've done just that.

Our farm sits on just two and a quarter acres about six miles from town. A little stucco cottage built in 1909, it's nestled above a reservoir lake and completely surrounded by orchards. A small kitchen garden, or *potager*, lies right outside the house and is currently filled with rows of egg-plant, tomatoes, peppers, Swiss chard, and green beans. The family cow, Cecelia, and our small herd of Katahdin sheep graze the pastures I can see from my bedroom window. A small flock of Pomeranian geese honks at the tractors they hear off in the distance. There are no less than a dozen bunnies currently grazing our lawn, fattening up on clover, hairy vetch, and various weeds. The pigs have made a gigantic nest of straw in the corner of their pen and are currently sprawled out, soaking up the breeze. And the three oldest children are running around, weaving in and out of the gardens. Here's hoping they at least have a few items of clothing on in case a neighbor happens to pay a visit.

The Reality

It's not by the sun that I can tell it's morning. And surely it's not by my desire to get out of bed. Rather, it's by the romantic crowing of my rooster (and yes, I would most certainly argue that a rooster's crow is romantic). It's the quintessential sound of a farm, and as it rings through the air, I know that my chickens are happily beginning their morning routine of wandering up to the house from their coop. They'll spend the next few hours scratching and pecking around in the grass and under rocks to find their breakfast. I always welcome them up to the house, awaiting their companionship as I begin my morning chores of filling slop buckets and milking the cow.

Chickens were one of the first animals that we added to our small farm (besides the Craigslist cow, obviously) because isn't there just some-thing so "farm-esque" about them? Their textured feathers. The gangster way they swagger around in the dirt while they hunt insects. Their various colors, shapes, sizes, and sounds. When one pictures a farm, it almost always includes baskets of fresh eggs, and thus I welcomed hens to our farm with great pleasure. I am a farmgirl, after all.

At first, I prided myself on the fact that they lived cage free. After all, that's what a "romantic" farmer would do. But one morning, I found them causing quite a commotion over some tender greens that had been pushing up from the soil in the late spring sunlight. These succulent, green tendrils were the Tommy Toe tomatoes I'd been nurturing for nearly twelve weeks indoors. I'd moved large pots of them outside only to find them tipped over, their roots scratched free of the soil, while the chickens pecked at the fresh foliage. The stems were broken, the leaves were crushed, many of the pots were broken, and the incident was later deemed "The Great Tommy Toe Disaster."

At that moment, I hated chickens, especially free-range ones. Who would ever invite such destructive creatures onto their property? Surely not me.

Welcome to the Farm

Many of us hold an ideal in our mind's eye of what backyard farming entails. For me, this includes free-ranging "chicken gangsters" (as I've come to call them), baskets of bountiful produce fresh and warm from the garden, old fence posts and rusty metal watering cans, morning dew, nests of multicolored eggs, the gentle call of a hungry lamb, flowering hops, mossy terra cotta pots, and nutritious suppers. I'm happy to report that many of these things do exist on a farm, and I'm ever humbled when I get to experience the romantic side of being a farmer. This lifestyle, even on the smallest scale, can be downright delicious.

And yet the reality is that this lifestyle can also be painstakingly tedious and difficult. Farm life isn't clean, organized, or easy. It involves an ever-present unpredictability, from the health of animals to the weather, which consistently challenges a farmer's sanity. Who in their right mind would choose a hobby, a passion, a lifestyle that would push them to the brink of an emotional breakdown? I'll tell you exactly who . . . a visionary. A dreamer. A hard worker. Someone willing to break free of society's molds about how we should fill our time and spend our money. Someone who is passionate about putting effort into the future of

her health and her environment. Someone who yearns to feel a connection with the soil, the winds, the seasons.

Though it's a simple way of life, it is far from being simplistic. Challenges arise, disaster strikes, and yet we still show up each day hungry for more. We are farmers.

Throughout the following pages, you'll find knowledge I had to learn, experience, and fail at multiple times before I knew enough to document the success. Trial and error is surely the mantra of the farmer! Use the guidelines, wisdom, and encouragement that fill these pages to build your own farm, in whatever capacity works best for you. For some, this will be a few pots of vegetables on the backyard patio. For others, it may be a small flock of chickens, or even a dairy animal. There is no right or wrong way to "farmgirl," so pick and choose from the pages what you will, or go all in! It's your dream, after all. And it's been waiting for you.

Welcome to the farm.

The Home Garden

For most of us backyard farmers, gardening is the gateway. Whether it's a pot of herbs on your front porch or a rented space in a community garden, gardening continues to be the foundation of the backyard life. With a load of dirt and a dream, it's easy to get started and requires only a bit of sweat equity on your part. It's a basic process that will only require a patch of soil, a sunny location on your property, and a few seeds to get started.

As you gain experience and knowledge in gardening, you'll most likely continue to expand your garden year after year. Most gardeners begin by planting a few of their favorite summer vegetables. Cucumbers, tomatoes, and peppers are all very common among gardening novices! These can easily be grown in a small raised bed—or even just in a pile of dirt. All that's important in beginning a garden is the desire to be connected with your produce. Once that desire bites, the methods and options are as numerous as the stars in the sky.

I often refer to my garden as "the potager." *Potager* is the French word for kitchen garden, but it's so much more than this. It can be full of produce, a variety of flowers, perennial herbs, architectural interest, and aesthetic appeal. A potager garden is designed to be as pleasing to the eye as it is satisfying to the belly. This is a wonderful goal for the backyard farmer who is growing for far more than just production. This is where we live—not just a place we visit to work. Drawing joy from the surroundings of our farm and garden is as important as drawing nutrients from its produce. Focus on creating a space that lifts your spirits and makes your soul sing. Our potager incorporates hand-split cedar posts that we commissioned from a local woodworker, an antique door that I fell in love with

years ago at a flea market, wrought-iron fencing and fence posts that I very enthusiastically took from a friend looking to move, no less than ten yards of pea gravel, three trailer loads of perennial flowers, cement posts, metal hanging baskets, repurposed antique planters, climbing arbors of honeysuckle, vintage light posts, and strings of lights. It's not all about maxing out your space. It's about creating a special space that makes you happy.

Benefits of the Home Garden

Naturally, the benefits of a home garden are just as numerous as the methods. Let's take a peek into just a few:

▷ *The freshest local produce available.* It doesn't get any more local than your own backyard! Because the harvest-to-plate timeline is so very short in the home garden, you're going to be eating even better food than money can buy. You know how fresh bread tastes right out of the oven? That's exactly how it is with the home garden. Freshness, particularly when it comes to produce, plays a big role in its taste. Once a vegetable is picked, it immediately begins to deteriorate. Home gardening enables you to easily put the freshest produce on your plate.

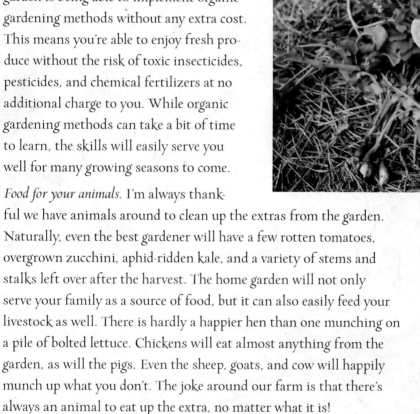

▷ *Organic without the cost.* Many folks steer away from organic produce at the market because of the potentially hefty price tag that comes with it. A big perk of the home garden is being able to implement organic gardening methods without any extra cost. This means you're able to enjoy fresh produce without the risk of toxic insecticides, pesticides, and chemical fertilizers at no additional charge to you. While organic gardening methods can take a bit of time to learn, the skills will easily serve you well for many growing seasons to come.

▷ *Food for your animals.* I'm always thankful we have animals around to clean up the extras from the garden. Naturally, even the best gardener will have a few rotten tomatoes, overgrown zucchini, aphid-ridden kale, and a variety of stems and stalks left over after the harvest. The home garden will not only serve your family as a source of food, but it can also easily feed your livestock as well. There is hardly a happier hen than one munching on a pile of bolted lettuce. Chickens will eat almost anything from the garden, as will the pigs. Even the sheep, goats, and cow will happily munch up what you don't. The joke around our farm is that there's always an animal to eat up the extra, no matter what it is!

▷ *Food for your compost pile.* If you don't have livestock to clean up your garden scraps, you can instead feed them to your compost pile! Your home garden will serve as a great contribution to the compost stack and will feed it many nutrients as it breaks down into that rich "black

Old bedding from the rabbit cage makes great, instant compost. It's got leftover stems from hay, poop, urine, and hair. All great additions to the garden. Pretty appetizing, isn't it?

gold" we're all familiar with. This compost can then be fed back to the garden as the cycle continues. It's pure magic!

Basics of Composting

Composting, by definition, is the decaying of organic matter to be utilized for plant fertilizer. There is no one way to compost, and each composter will have her own tricks of the trade that serve her home garden most fruitfully. We tend to keep the "lazy man's compost pile" around our farm, which would most likely make proper, organized composters shudder. Regardless, once you begin to reap the rewards of composting, you'll see it's worth every ounce of effort. Here are some basic composting methods:

Lazy Man's Compost

This is my preferred method because, well, I'm lazy. This method simply involves picking a spot on your property and piling all your garden waste,

kitchen scraps, grass clippings, yard waste, and animal manure/bedding there. You could calculate how much manure you pile on versus how much garden waste versus how much yard waste, if that's what you're into. Me? I'm more of a throw-it-all-in-a-pile-and-come-what-may sort of composter. But to each his own. Lazy Man's Compost will take longer to decompose, but with water and time will eventually get there!

Cow manure that's been left out in the elements for a while makes a wonderful and easy compost for the garden.

Trench Compost

A step up from the lazy method is trench compost, which simply involves digging a hole (or trench) in your garden soil and burying the garden waste, kitchen scraps, grass clippings, yard waste, and animal manure/bedding there, before covering it with soil. The waste will break down and will feed the soil in that particular area as it does so. All that's required on your part is digging the trench and avoiding planting anything directly over the compost trench for a few months while it decomposes and works its magic.

Open Bin Composting

Often thought of as the "proper" composting method because it's fancier (read: a bit more work), this form of composting will produce compost faster than the other methods listed, but requires more elbow grease and forethought on the gardener's part. Three open bins are set side by side in a desirable area on your property. The first bin is filled with the garden waste, kitchen scraps, grass clippings, yard waste, and animal manure/bedding ready to be composted. Once the items have begun to decompose, the compost is shoveled into the second bin and "turned" (think of this as stirring the compost) where it will be wetted down to maintain moisture. The compost will continue to decompose in this second bin before once again being turned and shoveled into the third bin, where it will finish its composting journey before being utilized in your home garden.

COMPOSTING TIPS

Moisture levels matter. Think about high humidity and what havoc that brings to homes! Water speeds the rate of decomposition. If you want your compost to break down faster, keep the pile moist! This is easy enough to manage with a hose or a few buckets.

Stirring increases activity. Each time compost is flipped—that is, the bottom part of the pile is moved to the top and the top is moved to the bottom—it aerates the waste and promotes decomposition. Without air, a compost pile will simply rot. No matter your chosen composting method, to speed along the process, turn the compost as often as you'd like to generate activity.

Get creative. The basics of composting are very simple and can be applied in a huge variety of ways. If none of the methods listed above with work for your situation, create one that will. Tumblers and prefabricated systems are always a great option.

Vermicompost

While it may seem slightly strange to keep a bin of worms under your kitchen sink, how much easier could composting get? In a vermicomposting system, red wriggler worms (*Eisenia fetida*) are placed in a bin. As the worms work through the scraps and mulch, they produce worm castings,

To set up a basic vermicompost system, you will need:

A large plastic bin or wooden box (5-gallon capacity or larger)
A few bricks to elevate the bin
Compost or peat moss
Sand, leaves, or yard mulch
1+ pound of red wriggler worms

1. Drill a few holes in the bottom of the plastic bin or wood box.
2. Fill the bottom of the box with 4 to 6 inches of peat moss. Throw a handful of sand over the peat moss (this helps the worms with digestion).
3. Gently pour the worms onto the peat moss.
4. Cover the worms with 4 inches of leaves or yard mulch. This will serve as the worms' bedding. Cover the bin with a lid to help keep the worms moist.
5. Feed the worms vegetable and fruit waste and garden waste as needed. Worms don't like processed foods any more than we do. Avoid them at all costs as they can cause the worms serious harm!
6. Don't feed the worms more than they can consume in a week. Feeding them in sections can help make this easy to identify! Feed the worms in one corner of the bin once a week. The following week, pour the waste into another corner, etc. Cover the waste with fresh compost and watch the worms work. You can easily harvest compost from the previous week's corner by scooping the compost out with your hands and filling the hole in with fresh waste and bedding.

which serve as a wonderful garden compost. In addition to the incredible compost, "compost tea" will drain out of the bottom of the system and can be diluted and utilized as a natural fertilizer. Vermicompost has been noted as one of the most balanced composts available. It is rich in calcium, magnesium, potassium, B vitamins, and phosphorus.

Urban Compost

Even if you live in a city apartment, composting is still totally within your reach. There is a large variety of small compost pails that will even look nice on your countertop. Pop open the lid, add in your morning eggshells, and find a local farmer or gardener who can utilize your scraps. If you're a real outlaw, you can even use some of them to feed your under-the-radar urban chickens or rabbits. Don't worry. I won't tell.

Basic Gardening Methods

I must admit, I'm a bit of a romantic when it comes to gardens. It's not enough for me that my gardens are productive. I want them to be beautiful too! What can I say? I'm a sucker for natural beauty. And there's hardly a better place to find that than in the garden. Luckily, with some very basic methods, your backyard garden can be both productive *and* beautiful.

As you begin to build your garden, take a few moments to step back and assess your situation. How much space do you have? How much time do you have to prepare your space? Maintain your space? This will help you establish the best gardening method for your situation and will serve you well as you get to work.

Square-Foot Gardening

If space is an issue, square-foot gardening is a fantastic approach. The principle is simple and easy to follow.

1. Divide the garden area into 12- by 12-foot sections. This can be in raised beds or simple dirt beds.

Tilling the garden is a great way to get lines set for row gardening. I make my husband do it because he's awesome and strong.

2. Plan and plant the garden (each seed according to its specific package directions).

3. See? I told you it was simple.

Pros: Those who appreciate efficiency and order make great square-foot gardeners. Because the garden is so well divided and planned, there is very rarely wasted space. Additionally, because the gardener knows exactly what is planted in each square foot, it's easy to plan for varying heights of plants, to plant companion plants close to one another, and to manage weeds.

Cons: The biggest drawback of the square-foot method is the intensity with which the plants are squeezed into the garden bed. Because this method maximizes garden production, it can also tend to be overplanted and not allow enough space for the plants to grow. This can result in poor root development and nutrient uptake from the overcrowded spaces. The square-foot method is great for smaller garden plots but is slightly unrealistic for a large garden bed.

Startup costs: Minimal.

Row Gardening

The most traditional form of gardening, this method utilizes simple rows to keep planning easy and production high. Most often, gardeners utilize this method when space is not as big of a concern. Row gardening tends to lend that traditional farm look to any garden, as the rows of produce grow into stunning lines of colors and flavors.

1. Till, amend, and prepare the ground for planting.

2. Place a stake at the end of each of the rows, so as to mark the length and width of each row.

3. Tie a piece of twine or rope between the stakes at each end of the row, which will serve as a line guide for planting. Use the corner of a hoe to dig a small furrow for planting, following each twine line. Sow seeds according to packet directions. Plant taller plants on the north side of the garden (or south side in the Southern Hemisphere), so as not to shade the shorter plants.

Early summer rows of crops can be weeded and mulched with ease.

Pros: I love how easy it is to plan out the garden with this method. The evenly spaced rows provide convenient access to harvests, as well as pathways to easily manage weeds. The method provides a clean and organized layout to the garden bed and can produce a huge harvest. It's the best way to utilize a large garden area.

Cons: Due to the layout of row gardening, it can easily result in a lot of wasted space if not managed well. Because the plants themselves don't create a "living mulch" (as with the densely planted square-foot gardening method), this method also runs the risk of exposing the roots of the crops if they are not properly mulched and managed.

Startup costs: Minimal.

Raised-Bed Gardening

An extremely popular way to garden, raised-bed gardening makes it so easy on us ol' ladies because bending over a garden bed is much easier when the garden bed is raised a foot off the ground! I loved that raised-bed gardening makes a home garden feel attainable and doable. Beds can be built out of lumber, pavers, or whatever your heart desires. A typical garden bed is about 3 by 6 feet and roughly 12 feet deep. Though every gardener will likely build their boxes differently, here's a simply method to building your own:

1. Make sure to have on hand a tape measure, pencil, carpenter's square, level, $2\frac{1}{2}$-inch exterior screws, drill, four 18- or 24-inch wood stakes, and four 2x6 boards cut to 8-foot lengths as well as four 2x6 boards cut to 4-foot lengths. Make sure the lumber has not been pressure treated, as this leaches chemicals into the soil.

2. After selecting a sunny location on your property, lay two long boards and two short boards on their sides in the shape of the garden bed.

3. Insert screws through the long boards, two per board, about $\frac{3}{4}$ inch in from the end of the board, drilling directly into the end of the short board.

4. Use the carpenter's square to check your corner angles. Then, insert a stake into each corner of the now-assembled garden bed, leaving about 6 inches sticking out above the garden box. Level the garden box and drill one more screw through the boards into the stake to help level and stabilize the box. You can repeat this with additional stakes along the interior of the box.

5. Last, set the additional 2x6 boards to create a second height layer to the box (giving the box a 12-inch depth). These can be assembled in the same way as the previous layer and can be screwed directly into the already-set stakes.

6. The garden bed can then be filled with compost and planted. Most people find that a few beds serve them well.

Pros: Raised beds are incredibly easy to weed, mulch, and harvest because of their accessibility. Raised beds allow for great drainage and offer protection from a variety of critters. They are also easy to mow around and look sharp if well maintained. They are a great option for those with limited space or those who wish to grow a smaller garden.

Cons: Building enough raised beds for a large harvest can be a bit daunting, as can the initial labor required to set up the beds. They're one of the most expensive options for the home gardener. Larger plants such as potatoes, corn, or melons also may find the boxes to be a bit restrictive for their growth. Additionally, as the lumber wears year after year, you'll need to face the music and replace your boxes.

Startup costs: Moderate. Depending on what materials you use to build your beds, or what you have lying around, the cost of raised-bed gardening can vary greatly. Recycled or repurposed lumber will obviously be the most frugal choice, whereas premade garden boxes will be the most expensive. Either way, filling the raised beds with compost can get quite pricey if you're not yet making your own. Buying compost by the pickup load will be a much more economical option than the bagged options.

Mint is a great candidate for container gardening (left), as it can be quite invasive in the garden bed! Old forgotten buckets and feeders (right) can make great containers for herbs, vegetables, and even strawberries. Even carrots can be grown in containers!

Container Gardening

Much like raised-bed gardening, container gardening simply utilizes a variety of containers to "contain" the garden. Get it? Contain? Container? All this involves is collecting containers, filling them with compost, and planting. Container gardening is a worthwhile option for those with very limited space or bad soil. Common containers include clay pots, plastic containers, whiskey barrels, concrete planters, metal buckets, and even foam pots. These can be picked up secondhand for a fraction of retail and can be as varying as you wish. I'm a hodgepodge person myself, but whatever suits your fancy.

Choose the right container for the right plant. Make sure that as you're planting your container garden, you opt for a container that will allow the plant to grow accordingly. Plants need lots of room to send

Top Ten Crops to Grow in Containers

Tomatoes

Herbs such as basil, rosemary, thyme, marjoram, oregano, tarragon, mint

Peppers

Greens such as lettuce, kale, collards, mustard greens, Swiss chard

Eggplant

Cucumbers

Carrots

Edible flowers

Zucchini

Strawberries

Beets

down their roots, so even if the plants are small to start, make sure you give them elbow room in their containers for optimal growth.

Pros: Perfect for small spaces, container gardening offers a huge bang for a buck. A small amount of space can yield even an apartment dweller a healthy harvest of homegrown produce. Container gardening allows for complete control of soil quality and pest control. Containers are easily managed, mulched, weeded, harvested, and moved around into optimum sunlight. They can even be easily brought inside during summer storms.

Cons: The biggest difficulty in container gardening is watering properly. If the pots are not properly draining, overwatering and its subsequent problems can easily become a huge issue. Alternatively, if the pots drain too quickly or the containers are particularly small, they will require much more frequent watering than others. A container gardener must be an attentive one.

Startup costs: Moderate. Obviously, to start a container garden, one needs containers. These can range in price from free to quite expensive, depending on the material and size used. On top of the containers, bagged potting soil can be a bit of an initial investment.

Maximizing Productivity with Succession Planting

The act of following one crop with another is an important concept for the home gardener to utilize. Why not get the most out of your garden? By implementing succession planting, it's easy to double (or even triple!) your garden's yield. The space in your garden serves double-duty time. That's what I'm talking about! Here's the basic method:

1. Make a list of all the crops you'd like to grow. Research these crops so that you know their individual requirements such as days to harvest, sun preference, space requirements, and frost sensitivity. This information can typically be found on the back of the seed packet.

2. Create a planting schedule. Making a planting schedule of what should be planted in the spring, summer, and fall will simplify the process. From there, you can draw out a diagram of your garden with what vegetables will be planted where during the different seasons. Some vegetables, such as radishes, will only be harvested for a few weeks during the spring or fall. Other vegetables, such as beets, can be harvested at varying sizes over the course of a few months. It would therefore be wise in your planting schedule to follow a short-season vegetable (such as radish) with a long-season vegetable (such as tomato).

3. Plant accordingly. On top of multiple crops from the same plot, it's also possible to extend your harvest and bump up production by simply planting more of the same crop at varying intervals. For example, instead of planting all your lettuce in one day, space it out over a few weeks. This means that you won't have to eat salad for every single meal once harvest hits. Instead, you'll be able to harvest crops at varying times as they come into maturity.

Cabbage is typically harvested in midsummer. Once it's gone, replant the space with greens for the fall.

Here are some examples of succession planting:

1. Plant carrots in between garlic rows. After the garlic is harvested in early summer, the carrots can continue to grow through the rest of the season.

2. Follow early-season tomatoes with arugula, radish, lettuce, or spinach for a fall harvest.

3. Spring peas can be followed by tomatoes or beans. By the time the tomatoes and beans can be safely planted, you'll already have been able to harvest lots of pea pods. After harvesting the tomatoes or beans, plant some fall arugula or spinach.

Seeds

Let's talk about seeds. Obviously, one of the foundational ways to ensure success in your backyard garden is to get the right seeds for the job. While the seeds available on the market are as numerous as grains of sand on the

beach, there are a few ways to determine which are the right ones for your garden. Spending some time during the winter months (preferably curled up next to a fire with a warm mug of chamomile tea) planning for the year's garden will really pay off. A wee bit of research in ordering your seeds will reward you big time.

Organic vs. Conventional

Organic seeds are seeds that have been grown by a parent plant on an organic farm. The farm must meet specific and stringent requirements in order to become certified as organic. Organic seeds are guaranteed not to have any seed treatments and often perform better in the garden. They are a common choice of the home gardener because they're the starting point for a garden free from synthetic pesticides and herbicides. A big perk of purchasing organic seeds is supporting farms and research that promote organic gardening methods. Think of it as a small investment in the future of our agricultural system.

Alternatively, conventional seeds are often treated with a variety of chemicals (most commonly synthetic fungicides and insecticides) prior to being sold. While they come with a slightly smaller price tag, conventional seeds are grown from parent plants in non-organic soil that could have been sprayed and treated with a variety of synthetic chemicals. If minimizing chemical exposure is important to you, avoid conventional seeds.

While a garden may start with organic seeds, this does not qualify it as an organic garden. In order to maintain an organic garden, one must avoid all synthetic fertilizers and pesticides. What's the use of organic seeds if you're going to spray them anyway? Likewise, conventional seeds can be used to grow vegetables in an organically managed garden.

Has your head exploded yet? But wait! There's more . . .

Heirloom Seeds

Heirloom seeds come from crops that have been selected and saved for generations. Perhaps they've been grown for taste, or performance, or color. Whatever the reason, the seeds from the very best produce are harvested, preserved, and handed down to be grown again. Despite popular belief, heirloom varieties tend to be very hardy (hence the reason they've grown and survived for so many generations!). I've always enjoyed the uniqueness that comes from heirloom seeds. The colors, shapes, tastes, and fragrances are endless. If superior taste is your number one goal, heirloom varieties should be on your list.

Hybrid Seeds

In the garden, it's common that plants are cross-pollinated—meaning the pollen of one plant is used to pollinate another. This is called open pollination and occurs naturally due to weather, insects, birds, and other factors. Alternatively, this cross-pollination can also be done by human intervention, resulting in a hybrid crop. First-generation hybridized crops experience "hybrid vigor," a term that basically means they grow bigger, faster, and stronger than their parent plants. Unfortunately, the seeds from these crops cannot be saved or preserved. A gardener who grows hybrid varieties must purchase new seed every year.

Genetically Modified Seeds

Genetically modified seeds are the result of genetic engineering in which the DNA from one organism is removed and put into the DNA of another organism. There is strong controversy to the long-term environmental and health effects of growing and eating genetically modified foods. Because genetically modified seeds are patented by the companies that engineer them, it is illegal to save seeds from genetically modified crops. As for me and my gardens? We avoid genetically modified seeds at all costs.

Selecting Varieties

After you've decided what type of seed you'd like to
grow in your home garden, it's time to figure out
which variety you'll grow! I know, I know—so many
decisions to make! But this is the fun part! Always
dream of purple beans? White cucumbers? Early-
ripening tomato plants? Now's the time to get wild!
Now that you've planned your garden, know which
crops you'd like to grow, and know which type of seed
you're looking for, you can select specific varieties of
that crop that'll fit the bill.

A few traits to choose from:

▷ *Yields.* If getting the biggest yield is your priority, select varieties that are known for their high
production!

▷ *Colors.* Sometimes half the fun is the novelty of
what you can grow yourself! I always love to pick a few oddly colored
crops to grow (such as freckled tomatoes) just for fun.

▷ *Flavors.* The mack daddy of all selections, this is primarily how I
choose my varieties. One question: Which variety tastes the best?
That's the one I want on my supper table.

▷ *Storage.* In addition to flavor, storage life can be a big factor in selecting which varieties of crops you'll grow. I choose my garlic, potato,
and onion varieties specifically for their storage qualities (since these
are the crops that we . . . wait for it . . . store!).

▷ *Days to maturity.* For those who live in colder zones, this can be a huge
factor in choosing the proper varieties for your garden. Put simply,
the shorter the days to maturity, the faster you'll have a crop! If you're
squeezing in harvests before frost dates, days to maturity will play a
big role in your gardening decisions.

*We select Yellow of Parma
onions for their storage qualities.
They start out like blades of
grass but grow into magnificent
bulbs in no time.*

My Favorite Seed Companies

I've chosen to spend my gardening dollars supporting companies that have a passion for growing and preserving organic, heirloom, and open-pollinated varieties.

Seed Savers Exchange

Sustainable Seed Co.

Bakers Creek Heirloom Seeds

Johnny's Selected Seeds

Seeds of Change

Territorial Seeds

▷ *Disease resistance.* Particular varieties are known for their disease resistance, making them ideal for the home gardener. This is always a nice addition to any variety.

▷ *Vigor.* Some varieties are simply hardier and stronger than others. If there's a particular crop I tend to struggle with, I will choose a variety that is known for its hardiness and vigor to stack the cards in my favor.

Basics of Seed Starting

Why start seeds? Great question, my friend. One would start seeds if one needed to get a jump-start on the gardening season. Up here in the north, we need to start a few crops indoors if we ever hope to get a harvest before the first frost. Tomatoes, eggplant, peppers, and other long-season

Starting seeds indoors brings with it the hope of all that's to come in the summer harvest. Give 'em soil, water, and light. Magic will do the rest.

vegetables require more frost-free days than we have. So starting them indoors allows us to grow them for a while inside, move them outside to the garden beds when it's safe, and harvest the bounty before the frost arrives in the fall.

There are other benefits to starting seeds yourself, not the least of which is control over what you grow. Unlike the nurseries and home-improvement stores that carry only a few dozen varieties, there are (literally!) limitless options of seeds available to the home gardener. This means you have freedom to choose exactly the variety you'd like. On top of increased growing options, the cost of seeds is significantly lower than buying starts.

Tips for Seed Starting

▷ *Don't start too many.* Often, it's easy for us gardeners to get overzealous (points to self) and start too many seeds from too many different varieties. Keep it simple. Once you've mastered a few basics, expand your repertoire.

▷ *Water from the ground up.* Seedlings are extremely sensitive and watering them from overhead with a watering can may cause disruption in the airy potting soil and can also cause the soil to crust over a bit. Instead, plant the seeds in pots with drainage holes in the bottom and place the pots in a tray that can hold water. This way, the soil and seedlings can absorb water from below as they need it, maintaining a consistent moisture level and preventing any damage to the soil aeration or seedling.

▷ *Follow package directions.* Often the seed packet will tell you exactly when to start your seeds. These instructions are there for a reason. If started too soon, some seeds (such as cucumbers) will suffer. I know rules are meant to be broken . . . but try to control yourself a wee bit.

Seeds require very little from us humans. Warmth, light, and a bit of soil is all they ask of us.

Light

When you start your seeds indoors, it's essential to give them access to light. In the winter or early spring, this means (at the very least) putting them directly under a southern-exposure window for optimized daylight. Because the days are still short this time of year, I've found it extremely beneficial to provide the seedlings with supplemental light as well. If the seeds receive too few hours of daylight, they'll struggle toward the light, resulting in leggy seedlings with weak stems. To grow strong, vibrant seedlings, I provide mine with artificial light until the daylight and sun exposure increases. I use a homemade setup of shop lights and sunlight-spectrum bulbs to supplement my seedlings. Ideally, the lights sit only 2 to 3 inches above the seedlings. As the seedlings grow, lift the lights up higher.

Eggplants and peppers require a lot of time to produce fruit. For those of us in shorter growing seasons, starting indoors is a must if we're to get any fruit at all.

Heat

Seedlings also require warmer temperatures to let them know it's safe to germinate. All seeds germinate at varying temperatures, but for the most part, room temperature (around 70 degrees F) will be enough to encourage them to grow. If you're keeping your seedlings in a cold basement or barn, additional heat may be beneficial. I very rarely supplement my seedlings with additional heat, with the exception of peppers and eggplants (they like it hot so I utilize a seed-starting mat under the pots to make it a bit warmer).

Seed-Starting Mix

A high-quality seed-starting mix will help to retain moisture and allow for easy root growth.

I won't admit that I've learned this lesson the hard way (and I definitely won't admit I've learned it at least two or three times) but for the sake

Basic Seed-Starting Guide

Equipment:

Seed-starting mix
Containers with drainage holes
Trays
Grow lights
Seeds

Method:

1. Write up a seed schedule so you know when to start which seeds.

2. Clean your containers with hot, soapy water to avoid any contamination.

3. Fill your containers with seed-starting mix. Using your fingertip or a pencil, create holes in the mix for the seeds. Plant according to package directions and then gently brush the soil back over the hole to fill it.

4. Place the containers in a shallow tray of water.

5. Label containers with the type of seed and date planted. Don't think you'll remember what and when you planted. You won't.

6. Cover the containers with clear plastic, if desired, to help maintain a moist and warm environment. You can remove the plastic once the seeds have germinated.

7. Place the containers 2 to 3 inches below a grow light, leaving the light on 14 to 16 hours per day. Adjust the light as necessary to always maintain a 2- to 3-inch gap between the bulbs and the plants.

8. After the plants have developed their first true leaves, you're safe to give them their first fertilizer treatment. I like to use compost tea from my worms, diluted to one-quarter strength with water. Fish emulsion is another option (though a slightly stinkier one).

9. As the plants grow, you may need to transplant them to larger containers. Take caution to always handle the plant by its leaves and not by its fragile stem!

10. Before moving out to the garden, make sure to harden off the seedlings. This process involves moving the plants outside on warm, non-windy days for a few hours and then moving them back in at night when temperatures drop. Do this for a few days, lengthening their time outside each day, so the plants can harden off and grow a bit tougher. Life in the garden ain't easy, baby.

of your seedlings, do not start your seeds in garden soil! Garden soil is alive. It's a teeny tiny little ecosystem that can actually put strain on your seedlings and affect their health and growth. Instead, grab a bag of organic seed-starting mix, which is gently moistened, light in texture, and free of contaminates. Potting soil makes it easy for your seedlings to take root and grow without any interference.

Water

Seedlings do best with consistent moisture. This means you can't drown them one day and then not water for a few days. Be consistent! Keeping the seedlings in a very shallow tray of water will keep you from having to stress about soil moisture, as the soil will naturally draw up exactly as much as it needs. Remove the containers from the tray once they've been thoroughly moistened. Your seedlings will not grow without water. Just in case you were wondering.

Herbs

One of the easiest ways to make your food tastier is to simply utilize the ol' kitchen herb garden. Yes, you can easily keep an herb garden in your backyard! Can't you just picture it? A beautiful hedge of rosemary, flowering chives, whimsical chamomile, fragrant lavender, and vibrant parsley. There's no shortage of flavorful–even medicinal–herbs that are easily available to the backyard gardener. For centuries, herbs have been concocted into tinctures and teas, used to flavor oils, vinegars, and butters, and in a variety of dishes. Having them available at all times in your garden is worth the effort!

Generally speaking, herbs can withstand average soil, though they prefer a fertile one. They also appreciate well-drained, sunny locations in your garden area. As with most plants, they benefit from consistent and even moisture, particularly while they're germinating, which can take anywhere from a few days to a few weeks. For those of us who are slightly less patient (that is, me), herb starts are usually available at garden nurseries

most of the year for only a few dollars each. My husband knows that one of my greatest weaknesses is when the herbs make their appearance at the local garden store in the spring. I'll happily keep buying flats of parsley, rosemary, sage, mint, and oregano until he starts to hide my debit card. I simply cannot say no to those beautiful rosemary plants. Confession: I currently have nine planted in my garden.

Because the entire purpose of growing herbs is for flavor and potency, it's important to keep the herbs from setting seeds or flowering. As flowers come, gently pinch them off. This will keep the herbs compact, flavorful, and growing!

My Top Fifteen Culinary Herbs

Basil

My love for basil (aka, "The King of Herbs") knows no bounds! I'm happy to devote a large section of the garden each year to this annual herb. Not only do we get to enjoy it fresh for months during the summer, but all extra basil is processed into pesto for the freezer. We're able to enjoy it atop pizzas and pastas throughout the winter. Basil is a wonderful companion plant for your tomatoes. I plant mine along the edges of my tomato patch. Pinching off the flowers as they appear will help keep the basil producing flavorful leaves and prevent it from getting leggy toward the end of the season.

Bay

I tend to use bay leaves most commonly with roasts and heavier soups in the winter. Ground up with garlic and salt, they are also a wonderful flavoring for chicken. The perennial shrubs can grow to be huge in favorable, mild climates (like, 8 to 10 feet huge!), so plan accordingly. Fresh bay leaves tend to have a milder flavor, while dried leaves pack a more powerful punch. Our bay shrubs must be planted in pots and rehomed inside during the winter–it's far too cold during the winters for their liking (this keeps me from having to restart with small ones each year).

Chives

One of most versatile herbs, chives lend a mild and tender oniony taste that enhances any dish. The best part? They reseed themselves and come back each year (it's best to contain them in some way if space is an issue). They'll spread willingly! I love the way their beautiful purple flowers shoot up in the summer. Chives are best when they are young and tender, so cutting back old, large spears is a great way to encourage new growth. Easily dehydrated or frozen, they continue to give throughout even the cold months, and they are one of the first to shoot up in the spring. They're a true workhorse in the culinary herb garden.

Dill

Because. Dill pickles. Am I right? I always grow a small patch of dill (also a great self-seeder) purely for the purpose of flavoring our pickled products. Green beans, asparagus, cucumbers, and beets are all great candidates for pickling and their taste is beautifully enhanced by dill's pungent fragrance. Even the flowers and seeds can be used!

Fennel

It took me a few seasons of growing it to fall in love with the taste of fennel, but once I did, I really fell hard for it! The flavor of fennel is reminiscent of a very mild licorice and lends itself well to summer vegetable salads and poached fish. Grown primarily as a biennial, both the feathery leaves and the bulb of the fennel can be eaten and enjoyed.

Garlic

Known as "Food of the Gods," garlic is invaluable in the kitchen. Garlic is the base of flavor in almost all savory dishes, just like its other close relatives in the *allium* family. We plant enough garlic to supply us year-round, concentrating on growing storage varieties that keep well throughout the winter. Garlic is grown by simply planting a single clove of garlic in the fall. In the spring it will begin sending up green shoots and garlic scapes and by midsummer, an entire head of garlic has developed below the soil line. Harvest garlic before it flowers by gently pulling it from the soil and allowing the heads to dry for a few days in a warm, dry, sunny location.

Lavender

What's that, Shaye? Lavender as a culinary herb? Oh yes! I've been known to sneak lavender into lemonade, cocktails, kombucha, teas, shortbread, pie crusts, meringues, and more! Lavender has a beautiful and gentle floral aroma that always takes me to a happy place. As far as I'm concerned, there's hardly anything more beautiful than a row of lavender. The flowers will bloom in the summer and can be harvested and dried for easy use throughout the year.

Marjoram

I've only been growing marjoram for a few years, but have already fallen in love with one of oregano's closest relatives. I commonly use it to flavor summer vegetable soups, egg dishes, and roast chickens. Call me crazy, but growing marjoram makes our home feel like a "proper" cottage, which is why we usually speak in a British accent when we harvest it.

Mint

For the love of two things, I grow *a lot* of mint: lamb roasts and mojitos. We raise and harvest lambs on our property and by far, my favorite preparations of lamb roast include a minty marinade, rub, or chutney. Its sharp fragrance cuts through fatty, red meat extremely well. And freshly picked mint muddled up in a cocktail? Well, that's what summer dreams are made of. Mint, a hardy perennial, grows vigorously and will eagerly (and quickly!) take over large sections of your garden. It's invasive, so plant with caution! Plant it along pathways, under trees, or in garden borders. Mint leaves are easily frozen or dried for winter use and with a bit of raw honey, make a wonderful tea.

Oregano is easy to grow and adds a wonderful flavor to pizza sauce and pasta dishes!

Oregano

The "pizza herb," as my daughter calls it, is a notorious contributor to a large variety of Italian dishes. Pizza, pasta, sausages, and chicken all benefit from oregano's earthly fragrance. Also known as wild marjoram, oregano is a staple for us, particularly in the summer garden when ratatouille and eggplant parmesan practically scream for it!

Parsley

The "duct tape" of herbs, in my opinion, is the ever-present parsley. Often grown as an annual, parsley will eagerly grow and produce lush leaves for harvest throughout the spring, summer, and fall. There is simply no dish that a bit of minced parsley can't enhance. It's tasty and beautiful. You can't beat it! I will often harvest a large handful of leaves at a time and keep them in a cup of water by the kitchen stove for easy access because parsley is used so often in our cooking. Curly parsley and

flat-leaf parsley are both herbs that will always, and forever, remain in our kitchen garden.

Rosemary

Rosemary is my favorite herb! But don't tell the others (though I'm sure they know already, by the amount of rosemary topiaries found throughout our potager garden). Because rosemary is a tender perennial, I dig mine up each fall and bring it inside for the winter. Though I try to control my usage while it's indoors, I can't help but sneak a few leaves into my morning eggs or beef stew. It's just too good! When we lived in Alabama where the weather was extra warm, rosemary was grown as hedges! What a dream!

Sage

Because we breed and harvest pigs on our farm, it's a necessity to keep sage growing in the garden. When we grind and mix up our sausage for the year, sage is a natural and welcome addition. An evergreen perennial, sage will continue to come back bigger and stronger every year. Its leaves range from a greenish-silver to a vibrant yellow-green. The fragrance of some sage is even reminiscent of a pineapple! There are many varieties to explore, all of which offer wonderful seasoning for meats.

Tarragon

When I began to cook through a few French cookbooks a few years back, I was introduced to tarragon, and knew it would need to have a home in our garden. I most commonly use it (along with a lot of butter) to flavor fish and chicken dishes. Tarragon is a perennial and can be harvested all throughout the summer. Make sure you source French tarragon, as it is the most flavorful of the tarragon varieties available. Then, channel your inner French chef, and get cooking!

Thyme comes back eagerly every year and is perfect for sprinkling over fried eggs at breakfast. Bonus: When it flowers, it provides a wonderful source of food for your bees!

Thyme

There's something romantic and dreamy about the evergreen herb thyme. The way it crawls along the garden beds with its teeny tiny, tender leaves is as aesthetically appealing as the way it tastes. Its fragrance is soft and mellow, yet distinct. Commonly grown as a perennial, thyme is an easy keeper in the kitchen garden. Sprinkle a bit of fresh lemon thyme over a homemade omelet, and you'll never think of breakfast the same way.

Fresh Usage

Throughout the spring, summer, and fall, almost all herbs are available for fresh use. This means you can waddle out to your garden in the wee hours of the morning with your basket and fill it with fragrant clutches of rosemary, parsley, and thyme for your breakfast casserole. You can enjoy large basil leaves atop freshly sliced tomatoes, and oregano in your homemade pizza crust. Does it get any better? Fresh herbs tend to be less potent than dried herbs, so use them freely and without restriction while you can!

Here are some helpful tips:

▷ *Wash your herbs in cold water before using.* Many herbs grow down low, close to the soil, and thus can easily get dirty. Offering them a quick rinse in a bowl of cold water will help to dislodge any stray bits of dirt or garden surprises that you may have harvested along with the herbs.

▷ *If you harvest more herbs than you'll eat right away, keep them in the refrigerator.* While it may be tempting to store all your herbs in jars of water, the stems will quickly get all sludgy and gross. To keep them as fresh as possible, simply wrap them up in a damp paper towel and tuck them into a large plastic bag before refrigerating until needed.

▷ *If the herb has a woody stem, pick the leaves and discard the stem.* If the herb has a tender, green stem, chop it up for extra flavor! Parsley and cilantro have beautifully fragrant stems that shouldn't go to waste. When finely minced, they make a wonderful addition to any dish along with the leaves. Woody thyme stems, on the other hand, aren't welcome at this party.

▷ *Got a ton of extra herbs? Put together some herb bundles for the freezer!* Soups are common meals in the cold, dreary winter months and herbs make a nice addition to them. Tying up small bundles of your extra herbs

Herb Bouquets

3 large stalks of parsley
1 large stalk of rosemary
1 large stalk of tarragon
2 large stalks of thyme

1. Cut all the stalks to 6 inches in length. Gather them together.
2. Use a 12-inch piece of kitchen twine to tie the bundle together at the top, in the middle, and at the bottom.
3. Place in a large plastic bag in the freezer until needed.

and putting them in the freezer is an easy way to preserve the herbs for use. I tie up my herb bundles with kitchen twine and store them in a gallon-sized plastic bag in the freezer. Whenever a soup needs a bit of flavoring, all I've got to do is throw in one of the herb bouquets!

The Spring Garden

The long, cold, dark winter days are almost unbearable. Soil, don't you love me? Aren't you ready for me to come out and play? As the sun's rays begin to peek through the gray skies and wake up the earth, the gardener tends to lose all self-control. Gardeners love getting dirty. Someone hand me a rake! This girl's goin' in.

The spring garden is not for the faint of heart. Spring is the time when you're up at dawn watching the sky for signs for rain or sunshine. It's the time when your nail beds ache from all the dirt that's crammed underneath them. And it's the time of year when all your dreams, your ambitions, and your goals for the garden are finally put into play. All those months of studying seed catalogs and drooling over seed packages are finally about to pay off! This is the time to get on your knees and commit to the backbreaking work that is gardening.

Waking Up the Soil

After a long, solemn, stagnant winter, the soil always deserves a bit of extra love in the spring. Here are a few methods for "waking up" your soil before planting:

Till: This can be a wonderful way to fluff up the soil, get the ecosystem kicked into high gear, and aerate the soil. Tilling helps to break up the soil, add oxygen, and mix in any soil amendments as well as winter mulch and/or scraps left from the year before. Tilling can be done with a mechanical tiller (for larger areas) or a simple garden spade (for smaller areas).

Fertilize your soil: Heavy winter snow and rain can cause valuable nutrients to drain out of your garden beds. Combat this by mixing in compost in the spring! Added a few weeks before planting any seeds, the compost

If you've got chickens, why not put them to work? Chickens are great at tilling up soil and fertilizing it at the same time!

will have time to mix in and spread into the soil without much effort on your part.

Test your soil pH: That is, if you're into that sort of thing. Some gardeners love the science behind gardening and a simple soil test helps them to know exactly what sort of soil amendments are necessary for their garden that year. Other gardeners, myself included, prefer to wing it–and then complain when things don't always turn out. You know. Because that makes total sense.

Sowing Seeds in the Garden

There are few times that I'm careful and meticulous on the farm. Sowing seeds (read: planting seeds) is one of those times. In my rebellious days, I'd throw caution to the wind, promise myself that I'd remember what

A large variety of lettuce can be directly sown into the garden early each spring.

and where I'd planted, and flippantly scatter my seeds about the bed. While this casual approach may still work well for some, when one is trying to maximize space and productivity, it's worth taking a bit (okay a lot!) of care when sowing seeds. Here are a few methods for sowing seeds:

Container sowing: Remember when we talked about starting seeds? In doing so, we also learned a valuable skill: container sowing, which is the art of sowing seeds in a container. This is primarily done with seeds that require a longer growing season than is provided by outdoor temperatures.

Direct sowing: This method involves sowing the seeds directly into the soil. To direct sow, all that's required is a tool of sorts (preferably a spade) to dig a shallow trench in the soil. The seeds are then planted to a particular depth, according to package directions, and covered with loose soil. After gently tamping down the soil and lightly watering it, you're all set!

Mechanical sowing: For the backyard farmer, this primarily means utilizing a seed planter. A simple and inexpensive device, the seed planter makes it extremely easy to plant even rows of various crops. Your seeds are

Cool weather-loving peas are easy to grow from direct sowing and will be one of the first to produce for you each spring.

WHAT BELONGS IN THE SPRING GARDEN

The spring garden beckons us in with its charm. Dainty flowers poke up through the soil, baby leaves start to spring up, and the smell of soil intoxicates the air. Sound romantic? It is. Spring is the time of beginning, of possibility. While your specific spring plantings will depend on your zone, the mission is broad and fairly general: Get to planting the following crops!

Peas	Radish	Lettuce	Cabbage	Collards
Spinach	Endive	Kale	Mustard greens	Broccoli
Brussels sprouts	Leeks	Rutabaga	Turnips	Green onions
Parsley	Beets	Carrots	Celery	Cauliflower
Potatoes	Swiss chard			

placed into a simple hopper that drops them at even intervals as you push the planter in a forward motion. This device will save your lower back–and those grandma knees.

Broadcasting: This is simply a term meaning, "evenly and gently scattering the seeds." This works best when you're planting a large area with a single variety of seed, such as wildflowers, grass, or wheat. The seeds are then covered with the appropriate thickness of soil and gently watered.

The Summer Garden

The summer garden will test you, push you, and challenge you. Pests begin to make their presence known, dreadfully hot weather can begin to take its toll, and the gardener (points to self) can easily become exhausted during the harvest! I've been known to throw a zucchini or two to my pigs instead

The general rule is to plant seeds in soil twice as deep as the height of the seed.

A gardener with a basket full of summer produce is a happy gardener indeed.

of preserving them because, well, I'm tuckered out! But regardless of the labor involved, the summer garden is always worth the work. As you work, you'll notice your plate full of organic produce, your larder and freezer filling up with preserved foodstuff, and your heart beaming with joy.

One of the greatest challenges of the summer garden is managing weeds. Weeds are the worst, man. And despite the fact that I can feed them to my sheep and chickens, I still resent their existence. I want to pull ripe tomatoes from the vine, not stupid morning glory from my potato patch. But weeds are a fact of life for the gardener, and no matter what type of gardening you do, you've still got to find a way to manage them well.

Natural Methods for Weed Control

Mulch

I put this at the top of the list because, frankly, it's my favorite method. Mulch gardening has completely revolutionized the way that I manage weeds. Gardeners use a variety of products as mulch (wood chips, straw, grass clippings, newspaper, feathers, etc.) but the premise is simple: cover the exposed soil with a thick layer of *something*. This barrier keeps the weeds from poking through. Not only that, but mulch holds moisture in the soil. I utilize aged straw in my gardens. I'll buy some bales, let them sit out in the elements for a few months to germinate any existing weed seeds in the straw, and then spread them all around my gardens, taking care, of course, not to cover the crops themselves. This keeps my soil at a consistent moisture level, protects the roots from harsh environmental elements, and produces a barrier for weeds. If weeds do poke through the mulch, which they inevitably will, I can pull them out like warm butter.

Homemade Sprays

Every gardener has a go-to homemade spray to combat weeds using anything from chili powder to vinegar. I keep a bottle on hand during the growing season but have found for larger gardens, it's not very much fun to squirt every weed. Makes the ol' hand tired. Regardless, it's helpful to have for walkways and edges!

Lay it on thick, baby! The soil can handle a lot of mulch, so why not give it a thick ol' layer? I apply a fresh layer every few weeks during peak growing season when the sun is the hottest and the weeds are at their worst, ranging from 4 to 12 inches deep.

Basic Homemade Weed Spray Recipe

1 gallon distilled white vinegar (5 percent acidity)

1 cup salt

1 tablespoon dish soap

Mix all ingredients together in a large bucket. Pour into squirt bottles. Spray on weeds, as needed (reapply after rain or watering). The weeds will wither up and die within a few days. Note: This weed spray is nonselective, so don't spray it close to your prized roses or luscious eggplants.

Animals

Yes, animals are a great method of weed control. Pigs do a heck of a job turning up the garden bed and eating everything in sight, weeds included! Ducks and geese also do a wonderful job of mowing the weeds, albeit less aggressively than the pigs. Many times throughout the year, I'll turn my laying hens loose in my garden beds. They'll happily spend their days scratching up weeds (and plants too) so I only use this method when nothing is growing or when it's all extremely established and can take a beatin' from the hens.

Hand Pulling

How shall I explain this method? One pulls the weeds. With her hands. One bends over and pulls the weeds. One kneels down and pulls the weeds. One digs out the weeds with a wee little hand shovel. One, by any means necessary, pulls the weeds from the soil so that they shall never see

the light of day again. I'll be honest: This is extremely satisfying for me. Someone grab my bucket and gloves. I've got work to do! I feed the freshly pulled weeds to the rabbits, sheep, chickens, or pigs. They get a snack. I get a free therapy session.

Fire

Fire kills things. Weeds included. If you need to clear a large area, and grazing animals isn't an option, using a torch to burn the weeds is a great way to kill those suckers. This method of weed control goes back thousands of years and, frankly, makes you feel like a pretty hardcore gardener. Added bonus: The ash from the burnt weeds feeds the soil. But wait! Make sure there are no fire bans in your area before getting busy.

Pest Control

Pest management can be a beast in the summer garden. While the vegetables are at their peak production, so are the pests! On the one hand, it's good for the home gardener to recognize that pests are simply part of the business. On the other hand, it's good to have some home remedies in your tool belt for when it's time to rein in their activity. When I first began gardening, I got squeamish at the idea of squishing a cabbage worm or squash bug. These days, I put on my big-girl pants and prepare for battle. Because most home gardeners don't choose to spray their garden beds with synthetic insecticides, it's essential to have a list of go-to methods.

Natural Methods for Pest Control

Diatomaceous Earth (DE)

This is a fine, whitish-gray powder that is made from the ground-up fossils of diatoms (that is, a particular type of algae). I'm no biologist, but I do understand the basics. Bugs crawl around in the diatomaceous earth, which cuts up and compromises their exoskeletons, essentially dehydrating them. Diatomaceous earth is commonly used in the storage of grains to prevent insects from contaminating the grain and is safe for mammals to eat. Did

Boiling water burns weeds too. So heat up that teakettle, pour the water over weeds, and watch 'em die. Don't use this method too close to crops, as it could affect their roots as well.

I mention it's cheap? And non-toxic? To use diatomaceous earth for pest control, simply sprinkle it on and around the plants where pests hang out. They will naturally crawl around in it and die. The tip to using DE is to sprinkle consistently. Any amount of water will render the DE worthless, so a fresh sprinkling after watering, rain, or a heavy dew is necessary.

Birds

Guess what birds eat? No, really, guess! Bugs! Therefore, it would stand to reason that the more birds you can attract to your garden, the more bugs they will eat from your garden. Feeders, nesting boxes, and bird baths are all great ways to attract our flying friends. If poultry is more your speed, bring in a few of your laying hens once the garden is established, and allow them to free range among the garden rows. While it may cost you a few tomatoes, they'll earn their keep quickly by managing bugs (and even give you a few eggs!).

Row Covers

This is one of the very most effective forms of pest management, particularly for squash bugs and cabbage worms, two of my mortal enemies! A row cover is simply a lightweight, transparent, tightly woven material that prevents a variety of pests from flying in and setting up shop in your crops. It serves as a block between the insect and your prized vegetables. Row cover is also effective in protecting plants from cold weather. Drape the row cover over the plants, either individually or over the entire row, making sure to secure the edges. Simply remove the row cover to harvest. Most row covers are water-permeable, so watering should not be a problem.

Row covers come in a variety of weights and sizes. Lightweight covers allow for very little heat retention and high light absorption, perfect for the summer months as a pest prevention!

Companion Planting

Though not the most effective method, it's common knowledge that certain garden plants fare best when planted next to others particular plants. Tomatoes and basil are a common combination, as the basil deters particular pests from the tomato plants. Companion planting may not do away with your pest problem, but it will most certainly help.

Common Companion-Planting Combinations

Corn with squash and beans
Tomatoes with basil
Garlic with beets
Carrots with beans
Lettuce with mint
Peppers with onions

Neem Oil

This vegetable oil is pressed from the seeds and fruit of the neem tree, an evergreen tree. When used properly, this oil is a wonderful all-natural pesticide in the home garden, is non-toxic, is biodegradable, and is widely available at garden supply stores. To use, simply dilute the neem oil with water in a spray bottle to recommended potency and spray on both sides of the foliage.

Hand Picking

I know, I know. Gross, right? But extremely effective. Every time I see the edges of my tomato leaves missing, I instantly go inspecting for the hornworm somewhere in the tomato jungle. Typically, if I follow the trail of bitten leaves, I can find him in no time. A quick grab and toss into the bucket means a snack for my chickens and no more tomato losses for me.

Although not really effective for aphids and such, this is a great method for some of the bigger pests. Still gross, however.

Crop Rotation

By simply growing your crops in different places each year, you can easily throw off the reproductive cycle of some of those bugs that lay their eggs in the soil, as well as fend off crop-specific diseases. When the bugs come back next year to feast on your squash, they won't be able to find them as easily!

Insecticidal Soap

This is not to be confused with dish soap, which contains a variety of synthetic fragrances and dyes. Approved for organic use, insecticidal soaps are made with specific fatty acids that are designed to dehydrate and kill bugs. Soap has been used as a natural insecticide for years. Sprayed onto the bugs directly, it can be highly effective in killing those super-pesky little guys, like aphids and whiteflies. The insecticidal soap is simply diluted with water to the recommended dosage and sprayed heavily on the infested foliage. Good contact with the bugs is necessary to be effective, and the treatment is best repeated every five to seven days as new eggs hatch.

Collars

A collar, placed around the stem of the growing plant to protect it from being eaten, is most effective for cutworms. I've found that a large, thick ring of aluminum foil is highly effective in preventing cutworms from munching on my spring goodies, such as broccoli and kale. When I transplant out the plants in the early spring, I place a small foil collar around the base of each plant. It's easy enough to remove once the plant becomes established, and is invaluable in those first few months of spring when *everything* is looking for something fresh and green to eat! Collars can also easily be made out of plastic cups, cardboard, or even toilet paper rolls.

Common Pests and How to Manage Them

All of the following should be repeated every few days until the bugs are under control.

Squash Bugs

Signs: Droopy leaves, black leaves, little to no fruit production, plant death.

Treatment: Row cover, diatomaceous earth, hand picking, insecticidal soap.

Cabbage Worms

Signs: Holes in the leaves and heads of brassica family members (such as cabbages, kale, cauliflower, or broccoli), green goo along the leaves.

Treatment: Row cover, diatomaceous earth, hand picking.

Slugs and Snails

Signs: Slug trails, nibbled produce, missing plants.

Treatment: Bait, hand picking, predators (chickens, ducks, and geese love them), copper collars, and diatomaceous earth.

Tomato Hornworm

Signs: Missing tomato leaves, withering plants, dark droppings on the tomato's leaves.

Treatment: Hand picking, tilling the soil after harvest to expose pupae in the soil.

Cutworms

Signs: Dead plants cut off at the base, entire plants missing (they're especially brutal in the early spring when plants are small and tender).

Treatment: Hand picking at night with a flashlight, collars, diatomaceous earth.

Potato Beetle

Signs: Clusters of bright yellow/orange eggs on the underside of leaves, eaten leaves, adult beetles on foliage.

Treatment: Hand picking, removing leaves with eggs, neem oil.

A baby cabbage worm is easy enough to squish between your fingers. Gross, I know.

You can tell a tomato hornworm by the spike on its end and the gigantic, dark droppings on leaves.

Catching and treating aphids before they really take over is the key to beating them

Japanese Beetle

Signs: Skeletonized leaves.

Treatment: Neem oil, hand picking.

Grasshoppers

Signs: Chewed leaves, holes in leaves, damaged fruit.

Treatment: Diatomaceous earth, insecticidal soap.

Whiteflies

Signs: Yellowed leaves, sticky honeydew over the plant, eggs on the underside of leaves.

Treatment: Insecticidal soap, removing leaves with eggs.

Aphids

Signs: White cast skins left behind, clusters of teeny little green bugs (especially on new growth), curled leaves, and sticky honeydew over the plant.

Treatment: Hand picking infested leaves, neem oil, insecticidal soap, blasting off with water. Ladybugs can also be purchased and released in your garden for wonderful aphid control.

Harvest Management

One of the biggest tasks for the gardener in those warm summer months is harvest management. Depending on the size and scale of your garden, it can become a *huge* deal. After all, what good is a perfectly grown tomato only to have it rot on the vine?

▷ *Harvest often.* One of the best ways to manage your produce is simply to harvest it often. During the hot months, when vegetable production is at its prime, I harvest every single morning. Basket in hand, I'll trudge out in the early-morning hours when the air is still brisk and

WHAT BELONGS IN THE SUMMER GARDEN

The summer garden is what dreams are made of, offering up baskets of bounty, rich foliage and flowers, and the bulk of all good things available to us from the soil. Much of summer's offerings come from what was planted earlier in the spring, so in many ways, your harvest is dependent on the success of your earlier plantings. That being said, there are still lots of crops to plant this time of year.

Basil	Beans	Corn	Cucumbers
Eggplants	Melon	Okra	Peppers
Pumpkins	Squash	Tomatoes	

HOW TO BUILD A TOMATO CAGE

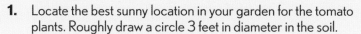

You will need:

Twine

Thick bamboo stakes, 5 to 7 feet in length

1. Locate the best sunny location in your garden for the tomato plants. Roughly draw a circle 3 feet in diameter in the soil.

2. Select five bamboo stakes. Drive each stake into the soil at least 8 inches at a slight angle so that they point inward toward the middle of the circle just slightly. The points should all be facing inwards, resembling a teepee of sorts.

3. Use a large piece of twine to bind the bamboo stakes together in the middle. This is where the support of the structure is, so make sure to take time and do it well, wrapping each stake carefully and tightly. I like to start with two and wrap them together. Then I add the third and wrap that in. Then the fourth, and so on. Adding one at a time can help to ensure they all get bound tightly. I'd like to say there's a perfect method for this, but frankly I'm too lazy for that, so my cages quickly became a wrap-it-until-it-feels-tight kinda thing.

4. Plant a tomato plant at the bottom of each bamboo stake, or alternatively, plant one right in the middle. As the tomato plant grows, take time to prune off suckers and bind the main stem to the bamboo stake with small pieces of twine. The bamboo will create a support for the plant as it grows.

the sun is just barely peeking above the hills. Everything that's ripe gets picked and brought inside for cooking or preserving. This daily harvest prevents any produce from rotting on the vine. Since it's very common to miss produce among the leaves, a quick daily harvest helps to catch all before it's too far gone.

▷ *Harvest in the morning.* Often, plants that are stressed out or limp during the heat of the day will perk back up after a good night's rest.

Because of this, I always prefer to harvest in the morning hours before the sun has taken its toll on the produce (and on the gardener as well!).

▷ *Eat your veggies.* Now is the time of year for eating more garden goodness than you can bear. Get creative. Scramble some collards into your morning eggs. Tuck sliced tomatoes into your sandwich. Add fresh green beans to your rice pilaf. Eat. All. The. Vegetables! There are plenty of months in the winter for potatoes and meat. When you *eat* your harvest, you'll *appreciate* your harvest all the more.

▷ *Store up.* If you're only gathering four or five pickling cucumbers each morning, it'll take some time before you have enough to put up your own pickles. But that's okay! Most vegetables store extremely well in the refrigerator and are more than happy to wait their turn. I often will bag up my green beans or peas, store them in the refrigerator, and deal with them all in one fell swoop. If I had to deal with small amounts of a large number of crops each day, I'd easily burn myself out. Rather, storing up the vegetables until I have a larger amount to process makes it all easier.

Sunflowers, even those past their prime, are still a welcome addition to the fall garden.

The Fall Garden

While I still love the fall garden, I must admit, it's not my favorite. Everything is past its prime, damaged by sun, harvests, and insects, and is bearing the burden of a job well done through the growing season. It ain't pretty. But it still feels wonderfully magical. The fall garden offers us gardeners the opportunity to squeeze a bit more from the earth and enjoy fresh produce a wee bit longer into the colder months.

Cover Crops

If I'm too exhausted to grow crops in the fall, which admittedly has been known to happen during a few pregnancies, there are still some wonderful ways to put the

garden to work and give it a boost. After the stress of a growing season, many gardens benefit from the growth of a cover crop. Quite simply, a cover crop is a crop that one plants to, you know, cover the soil. Like a blanket for the earth, the cover crop will establish roots, protect and preserve the soil structure, and can even add nutrients right back into the soil, increasing its fertility. And that's a win-win-win, if you're counting.

Common Fall Cover Crops

- *Hairy vetch.* A legume, hairy vetch is ideal for adding nitrogen back into the soil. It grows quickly, prefers fertile soils, and will need to be tilled in in the spring. Plant 1 pound per 100 square feet and lightly cover with soil.

- *Field peas.* Peas are legumes, which means they have the ability to take nitrogen from the air and put it back into the soil. How radical is that? Plant 5 pounds per 100 square feet and lightly cover with soil. Peas are fast-growing and efficient at choking out other weeds.

- *Ryegrass.* This crop is fast-growing and easy to grow! It will go dormant over the winter and resume growth in the spring, so tilling it into the soil in the spring will be necessary before planting other crops. Plant 1 pound per 100 square feet and lightly cover with soil.

- *Winter wheat.* Wheat is fast-growing, prefers fertile soil, and will naturally die off in the winter. Plant 2 pounds per 100 square feet and lightly cover with soil.

- *Oats.* Will die off naturally in the winter, making it ideal for gardeners who want to work their gardens first thing in the spring. Plant 4 pounds of oats per 100 square feet and lightly cover with soil.

The Importance of Fall Mulching

If messing with cover crops ain't your thing, have no fear! There's always mulching. A nice, thick layer of mulch (think 6 to 12 inches or more) will protect the soil from the cold nights and harsh weather that comes with the season. It's insulation for your soil. Couple that with the fact that mulch is literally falling off the trees this time of year and the world makes total sense! Add all those beautiful leaves, needles, grasses, and hay to your empty garden bed with abandon. Lay it on thick, baby. The layer of mulch will protect the soil from erosion, suppress weeds, retain water, protect perennials from severe temperatures, and add nutrients back to the soil as the material decomposes.

My favorite garden mulch is straw hay and/or rotten hay. Most hay will contain seeds, which you won't want to introduce to your garden, so it's important that the hay has had a chance to sit out in the weather for a while. This allows the seeds in the hay to germinate and die before you introduce the mulch to your garden. Rotted hay is like gold around these parts!

Common Mulches

Pine straw
Rotted hay
Straw hay
Rotted manure
Leaves
Grass clippings
Wood chips
Cardboard
Newspaper

Utilizing Your Natural Tillers

Fall is a great time to send your natural tillers into the garden: your animals! They have a wonderful way of scratching and rooting up all sorts of leftover goodies from the soil and in return, will drop their poo where it can break down and fertilize the soil.

Chickens are a natural choice for tilling up your fall garden. They require minimal fencing to contain and are wonderful at scratching up bugs, fertilizing the soil, and eating up leftover produce and its seeds, which can be a real nuisance in the spring. There's hardly a better sight than watching the chickens take over work in the garden.

Though they require a heavier fence to keep them in, pigs are a fantastic way to quickly and easily till up the fall garden. They're ruthless at rooting up leftover potatoes, tomatoes, and those thick broccoli stalks that are almost impossible to pull out. Pigs do the heavy lifting so you don't have to! We like to fence off our garden and give them a week or so to work their magic before planting or mulching.

WHAT BELONGS IN THE FALL GARDEN

Surprisingly, there's quite a lot to be grown in your fall garden! Many vegetables thrive on the cooler temperatures that autumn brings and will continue to give a crop well into the frosty days. The secret to the fall garden is making sure you give the plants enough time to get established before the shorter, colder days set in. For most plants, this means getting them into the ground in late summer when they'll still have plenty of warmth and sunshine to get growin'. A few crops, like garlic and shallots, should also be planted in the fall for spring harvest the following year. They will lay dormant under the soil through the cold winter before shooting up in the early spring. They should be mulched heavily to protect them from the harsh weather.

Broccoli	Cilantro	Brussels sprouts	Spinach	Kale
Collards	Mustard	Lettuce	Cabbage	Cauliflower
Kohlrabi	Radish	Rutabaga	Turnips	
Carrots	Pumpkins	Leeks	Arugula	

As you're cleaning up the summer garden in preparation for the fall, make sure you save all the waste. Most stalks, leaves, spent produce, and stems can be eaten by chickens, pigs, sheep, geese, and rabbits. The exceptions to this are members of the nightshade family, such as tomatoes, potatoes, peppers, tomatillos, and eggplant, which can be poisonous to animals. These are best thrown in the compost pile.

The Winter Garden

While some gardeners in warmer zones are able to grow cold-season crops through the winter, around these cold parts, winter time is primarily rest time for this ol' gardener. And truth be told, I'm usually extremely thankful for the break! After animal tilling and mulching the garden in the late fall, I sit back and enjoy a job well done. I usually start seeds indoors in February, so the break is short lived, but welcome.

The Winter Garden To-Do List

Fill out your gardening journal

Which varieties were your favorites?

Which flavors did you like the best?

Did you have any total flops?

Make a plan for your spring garden

Select and order seeds

Fix and/or sharpen gardening tools

Dream!

Harvesting through the Winter

In mild climates, it's possible to keep some of your fall produce in the ground for winter harvest. Heavily mulched (and I do mean *heavily*) vegetables such as potatoes, carrots, beets, turnips, and rutabagas can be stored right in the soil! Harvesting can be a bit difficult in the snow, but this is a great way to utilize free cold storage. Kale, spinach, and collards, all of which tolerate frost well, can often be heavily mulched and harvested into the winter months as well.

Growing for Longer

A great way to lengthen your growing season, especially for those in short-seasoned areas, is to utilize a cold frame, hoop house, or greenhouse. These options will make it possible for you to grow produce in your garden

If you're not growing, someone else may be! Hit up your local produce market to see what sorts of goodies are filtering in. I can often find apples, mushrooms, hydroponic lettuce, root vegetables, a variety of sprouts, and greens all the way through the winter.

Leaves are a wonderful (and free!) mulch to pile onto the garden beds each fall.

for a longer period of time. More growing means more food. More food means more eating. And more eating means more happiness. Amen.

Mulch

One of the easiest ways to lengthen the growing season is to utilize mulch. Plants that are heavily mulched (think 6 to 12 inches) will be able to withstand cold better than those left unmulched. As the cooler weather sets in, build up a thick layer of mulch over the roots and stems of your plants. The mulch will help regulate water retention and temperature. The more mulch, the more protection. Root crops, such as potatoes and carrots, can also be stored in the ground over the winter without freezing if they're under a thick layer of mulch.

Row Cover

The least expensive of the bunch, row cover is often utilized to extend growing seasons by offering frost protection for the rows of produce already growing in your garden. Hence the name: row cover. Small hoops are pushed into the ground at even intervals over the row of plants. Row cover fabric, a finely meshed fabric, is then draped over the row and secured down with dirt or rocks. Row cover can be utilized to protect plants from frost, overheating, severe winds, the spread of disease, and even insects! Someone remind me to drape this over my cabbages next spring! It is not water permeable, so be sure to remove it while you're watering.

Cold Frames

Cold frames come in a variety of shapes and sizes, but the purpose is simple: to protect the plants from the cold. Typically, cold frames are essentially raised garden beds, built up higher on the sides, with hinged windows that can be closed over the bed. This is a great way to grow lettuces and greens earlier in the season, as well as to extend the life of your cold-sensitive plants in the fall. During the warm summer months when no protection is needed, the window can be flipped up and left open, making it easy to monitor and control the temperature. Cold frames are a great project to build from recycled material. We've built small stone beds to the specific size of windows from our kitchen renovation. Zero dollar investment and more homegrown food? Yes, please.

Check out what zone you live in to determine when to plant.

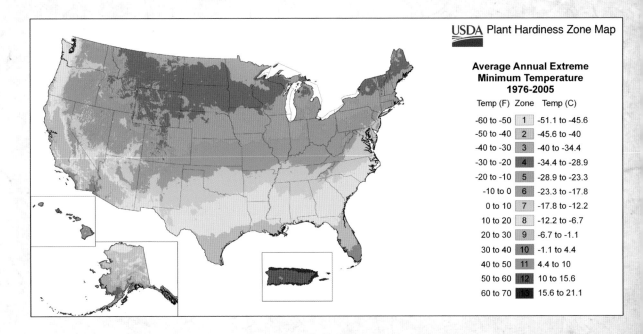

USDA Plant Hardiness Zone Map

Average Annual Extreme Minimum Temperature 1976-2005

Temp (F)	Zone	Temp (C)
-60 to -50	1	-51.1 to -45.6
-50 to -40	2	-45.6 to -40
-40 to -30	3	-40 to -34.4
-30 to -20	4	-34.4 to -28.9
-20 to -10	5	-28.9 to -23.3
-10 to 0	6	-23.3 to -17.8
0 to 10	7	-17.8 to -12.2
10 to 20	8	-12.2 to -6.7
20 to 30	9	-6.7 to -1.1
30 to 40	10	-1.1 to 4.4
40 to 50	11	4.4 to 10
50 to 60	12	10 to 15.6
60 to 70	13	15.6 to 21.1

Hoop Houses

Hoop houses are also a great way to extend the growing season. These are easy and inexpensive to build wherever you see fit in your garden or on your property. Hoop houses are typically temporary structures comprised of hoops made from metal, PVC pipe, or wood that are then covered with a thick plastic sheet. The hoop house is warmed by the sun and can change the interior climate by as much as two zones, generating more heat than just row cover alone! These can be built from scratch or purchased in a kit.

Greenhouses

I have a romantic affinity for greenhouses–especially the antique, European varieties. While I may not have one of these, I can still appreciate the aesthetic appeal of a well-built greenhouse and the welcome extended growing season it brings to the garden. I've never grown better peppers, eggplant, and okra outside than I have in our greenhouse where the plants are protected from wind, cold, rogue chickens, and roaming children. There are limitless options when it comes to building a greenhouse on your property, so take the time to plan out where you'll want it, what size it should be, and what style will fit your property best. Decide whether you want a vented greenhouse, or perhaps even one with a heater, so that you can easily regulate temperature year-round. There are prefabricated greenhouses available for purchase from a variety of feed stores or online retailers, though I'm much more drawn to the imperfect home-designed variety. Our first greenhouse was imperfect, no doubt, but it also displayed an antique door that we picked up at a thrift store for a bit of romance. Swoon!

I love greenhouses for the way they help me extend the growing season.

Seed Saving

Almost as old as the art of gardening itself is the art of seed saving. Each year, crops produce seeds that must be harvested, stored, and kept for the following year's garden. It's nature's way of making sure you've got exactly what you need to grow another harvest. While many people no longer bother saving seeds from their garden, it's a wonderful way to selectively

improve your crops each year. And did I mention it's free? After years
of saving seeds, you'll be left with a seed bank that directly reflects your
personal tastes as a gardener.

Saving the Right Type of Seeds

Seeds are best saved from crops that are non-hybridized and non-GMO.
Hybridized seeds rarely breed "true to form," so if seeds are stored from
your favorite hybrid varieties, you'll very rarely end up with the same plant
the following year. They're really only good for one generation. Genetically
modified seeds are protected by patents, meaning that it is illegal to
harvest, store, and replant them. But don't fear! All heirloom varieties can
easily be harvested and stored for next year. It's one of the many beneficial reasons to buy and grow heirloom vegetables. Organic vegetable seeds
labeled as such cannot be genetically modified. They're your best bet for
saving.

Ensuring Purity of Heirloom Varieties

If you want to preserve the genetic makeup of an heirloom variety, it's
important that the plant does not become cross contaminated by any
other varieties. There are a few simple ways to prevent contamination:

1. Place a paper bag over the individual flowers to ensure no pollen from
    the different varieties contaminates the heirloom seeds.

2. Only plant one variety at a time (i.e., one variety of eggplant).

3. Position your garden plot as far away as possible from your neighbor's
    gardens.

Plan Accordingly

Some vegetables are biennial, meaning they don't produce seeds until their
second year of life. This means they won't produce seeds until their second
year in your garden as well. If they're not hardy enough to survive through
the winter in your area, you will need to plan on purchasing new seeds
each year for those crops. Some biennial vegetables, such as carrots, are
very cold-hardy and can survive the winter in many zones. They will send
up flowers (and subsequently seeds) the following spring.

This year's dried beans will provide us with next year's crop!

Seed-Storing Tips

Though seeds don't require much of us, they do have a few basic requests. They like it cool. And they like it dry. Not too much to ask, is it?

Package: Paper envelopes and sealable jars are both great options for storing seeds. I prefer the envelopes because they take up less space and are less likely to break. Some gardeners like to add in a few pinches of powdered milk to help absorb any moisture.

Label: I know, I know, you think you'll remember, but trust me. You won't. Spend just a few moments writing down the variety included and the year the seed was harvested.

Store: A root cellar is a wonderful place to stash seeds (think dark and cool), but in a pinch, a refrigerator will work as well. Some seeds store better and longer than others, so it's important to keep track of what you've got in your vault. The average life of a stored seed can vary between two to ten years, depending on the variety and the exact conditions of the storage facility. It's important that you continue to grow the crops and harvest new seeds to keep your supply as fresh as possible.

How to Collect Seeds from Common Vegetables

Remember, only keep seeds from the healthiest and most vigorous plants in your garden!

Tomatoes, Cucumbers, and Melons

Cut the fruit in half. Squeeze or scoop the juices and seeds into a small jar or bowl, filling the container about half full. This will prevent the seeds from drying up before fermentation has completed. Place the container in a location where it can sit, undisturbed, for one to four days. Once the gelatinous seed coatings float to the surface, commonly along with a layer of white mold, the seeds are done fermenting. Add a bit of water to the container and swish it around. Mature seeds will sink to the bottom. Pour the contents through a fine, mesh strainer. Rinse under cool water to expose the mature seeds. Spread the seeds out on a paper plate to completely dry. This can take a few weeks. Label and store accordingly.

Peppers, Eggplant, Squash

Cut the fruit in half, remove the seeds with your fingers or by scraping them out with a spoon, and place on a paper plate to completely dry. Store accordingly.

Peas and Beans

Choose some of the best pods to leave on the plant after harvesting the rest. Allow the pods to dry on the plant before snipping them off, breaking open the dried pod, and picking the seeds out. Discard any pods that have become wet by rain or sprinklers. Spread the seeds on a paper plate and allow them to dry completely. Label and store accordingly.

Carrots (Biennial)

Allow the carrots to remain in the ground, unharvested, through the winter and into the spring. If they survive the winter, they will send up large, wispy, white flowers the following year that resemble Queen Anne's lace. Eventually, small seeds will form on the flower head that can be harvested by snipping the flower head from the stalk and shaking it over a large bucket, banging it on the side to dislodge any and all mature seeds. Separate the teeny seeds from any chaff from the flower head and store accordingly.

Lettuce

Lettuce must be left in the ground when the weather really begins to heat up in order to encourage "bolting," that is, beginning to grow a flower stalk instead of edible leaves. Once the lettuce bolts in the heat, the flower stalk (and its subsequent seeds) will begin to form. After the flower head matures, it will appear dry and slightly fluffy. Snip the flower heads from

Leave the best beans on the bush to let them dry before harvesting the seeds for next year's crop.

their stalks and shake them over a large bucket, banging them on the sides to dislodge any and all mature seeds. Separate the teeny seeds from any chaff from the flower head and store accordingly.

Kale, Collards, and Mustard Greens (Biennials)

Allow the greens to live in the ground over winter (this may require a hoop house or greenhouse depending on your zone). In the spring, allow the greens to send up flower stalks and go to seed. Snip the flower heads from their stalks and shake them over a large bucket, banging them on the sides to dislodge any and all mature seeds. Separate the seeds from any chaff and store accordingly.

Beets, Chard, Broccoli, Cauliflower, Cabbage, Kohlrabi, and Spinach (Biennials)

Allow the plants to live in the ground over winter (this may require a hoop house or greenhouse depending on your zone). In the spring, allow the plants to flower and go to seed. Once the flowers are brown, snip the flower heads from their stalks and shake them over a large bucket, banging them on the sides to dislodge any and all mature seeds. Separate the seeds from any chaff and store accordingly.

Corn

It's important for proper pollination and the growth of full ears that corn be planted in blocks. Allow the ears of corn to dry on the stalk. Harvest as soon as they are dry, remove the husks, and allow the cobs to dry completely. Once dry, rub the ears with your hand, over a bucket, to remove the seeds. Harvest corn kernels from as many ears as possible to encourage genetic diversity.

Germination Rates of Stored Seeds

The longer seeds are stored, the lower their germination rates become. Germination is when the seed sprouts into a plant. Longer storage equals decreased germination equals fewer plants. You still with me? Great. It's an important concept to remember if you're collecting and storing your

If you're having a hard time keeping biennial plants alive through the winter, it is possible to dig them up and plant them in pots before moving them into cool storage (ideal environment is around 45 degrees F and relatively high humidity) until spring, when they can be replanted.

own seeds and will even impact how you order seeds. If you only plan on planting a few tomato plants, it's probably not a good idea to order a gigantic package, right?

This also means that in order to keep your seed storage fresh and viable, you'll want to continually grow the plants and harvest fresh seed from the best specimens.

Basic Home Germination Testing

Using the seed package, or basic knowledge of the plant, note the requirements of each seed you'd like to test. For example, some seeds will require a warmer temperature, a grow light, or complete darkness in order to germinate.

1. Using a permanent marker, write the seed variety, date, and number of seeds that you'll be testing onto a paper towel.

2. Gently moisten a separate paper towel with a squirt bottle. Too much water will cause the seeds to mold. Space the seeds evenly around on the paper towel. Ten to twenty seeds should give you a pretty accurate germination rate.

3. Layer the paper towel with the written information over the top of the seeds. Squirt it with water as well to moisten the towel. You've created a paper towel and seed sandwich! Booya. Now, roll the paper towels into a sausage shape and seal in a plastic bag. Place it in an environment suitable for the specific seed's germination requirements.

4. Each day, check the seedlings to ensure the paper towels have remained moist, squirting them with water if necessary, and noting any changes in the seeds. Discard moldy seeds immediately.

5. When the normal germination time has passed, or all the seeds have sprouted, the test is over. Calculate your results. (Healthy seedlings / Total seeds) x 100 = Germination percentage.

Now that we've focused on growing all this delicious food, let's talk about how to preserve it so we can enjoy it for as long as possible!

CHAPTER 2

Preserving the Harvest

"Preserve everything!" is the homesteader's mantra during the months of plenty. You'll see many a farmgirl hunched over the stove canning her precious peaches, dehydrating her cherries in the sunshine, or bundling up herbs to dry. Traditionally, this is the only way people had food to eat during the winter. I suppose at least a tinge of that tradition has held on through the years. Be it through canning, freezing, dehydrating, culturing, or cold storage, preserving the bounty of spring, summer, and fall is the very best way to ensure you're eating well all year round!

Canning the Harvest

One of the most common ways to preserve a surplus of succulent summer goodness is by canning. Each year, homesteaders line up with their glass jars and zealous ambitions to fill their winter larders with the best of what the harvest has to offer. And after popping the top on those luscious, gently sweetened peaches while the snow falls, you'll totally understand why it's worth the effort. Summertime on our farm means boxes and boxes of ripe produce, sticky floors from "helpers" overfilling the jars, bottles of vinegar lined up along the walls, and the steam from the canner filling the kitchen. It's hardly a bad place to be.

Almost all vegetables, fruits, and meats can be preserved by using various methods of canning. What canned goods does your family currently enjoy? Are there specific crops that you'd like to invest your energy into preserving? We're surrounded by orchards in our area, so it's an annual ritual to put up a zillion (fine, it's closer to a million) jars of cherries, nectarines, peaches, and apricots. There's hardly a treat that makes my kiddos

Only use the best produce for canning. Bad fresh food will produce bad canned food, no matter how much sugar syrup you add!

Canned apricots are a wonderful winter treat drizzled with honey and sprinkled with walnuts.

giddier than a bowl of canned cherries. We also put a lot of our canning efforts toward pickles, because I seem to be eternally pregnant, and I can hardly get enough. Asparagus, beets, cucumbers, a variety of green beans, and even eggs can be very easily pickled. If pickles aren't your weakness, as they are mine, how about canned beans? Tomato sauce? Corn? Jams and jellies? Dream of the jars lined up in your storage pantry, and let's get to work.

There are two basic methods of canning: water canning and pressure canning. Both are wonderful methods for the backyard homesteader to delight in.

Water Canning

Water canning is a beautiful preservation method for your excess food-stuff. When I delight in my memories of homesteading as a young girl, this is often the task that comes to mind. My grandpa often canned peaches, pears, and applesauce throughout the summer and fall, and the smell of warmed apples still takes me back to his kitchen. I remember being down in his root cellar, grabbing a few jars of pears off the dusty shelves, and feeling rich as I marched proudly up the stairs with my bounty in hand. Grandpa used to sprinkle shredded cheese over the top of his pears, which still gags me to this day, but I'll happily dive into a bowl of plain, preserved pears any day!

The Basics

Bacteria require oxygen in order to grow. Thus, when oxygen is removed from food, bacteria are unable to do what they like to do. Water canning utilizes a large water bath to aid in removing oxygen from food, thereby preserving it. Food is packed into sterilized glass jars that are boiled in a large water bath for a designated amount of time. This renders the foodstuff shelf-stable for years to come! Water canners, typically made from aluminum, will most commonly hold up to seven quart-sized jars at one time, covering them completely with water. Water canning makes it possible to enjoy tomato sauce, a variety of pickled foods, chutneys, jams, jellies, and canned fruit all throughout the winter months when harvests have stalled or stopped completely.

The Catch

Water canning relies on the food's acidity for part of its effectiveness. Because of this, the pH is often altered to reach a particular acidity. For example, green beans cannot be water-canned without vinegar, as the acidity of the beans alone will not be enough to hold them in storage (they will spoil . . . and you'll know it). High-acid vegetables, such as tomatoes, do very well with water canning and almost never require additional acidity.

Hit up local producers to get deals on their surplus produce. When fruit is in season around our parts, we can usually find people who will either give it away for almost nothing or allow us to come pick for free! Look for signs, markets, and online listings that can direct you to producers in your area.

The Method

Water canning requires a few basic supplies that are available at most homesteading, feed, or hardware supply stores:

Water-bath canner   **Jar tray**

Mason jars   **Towel**

Metal lids and bands   **Wooden spoon**

Jar tongs

1. Fill the water canner with water and bring it to a simmer. When you add the jars to the canner they will add volume, so take note of where your water line will need to be once the pot is filled with jars.

2. Pack the foodstuff, according to your recipe, in sterilized glass jars (this could mean raw or cooked foodstuff, depending on your recipe).

3. Wipe off the rim of the jar to ensure a good seal.

4. Place the metal lid on the jar, seal side down, and twist on the metal band—but not too tight, now!

5. Place the jars carefully into the jar tray and lower slowly into the hot water bath. Remove extra water as needed so that the canner doesn't spill over! The water should cover the lids of the jars by 1 to 2 inches.

6. Put the lid on the canner, bring to a simmer, and process the jars for the amount of time designated by the recipe.

7. Carefully remove one jar at a time from the canner with the jar tongs and place on a tea towel to cool. Let the jars sit there for twenty-four hours. You'll hear them begin to *pop* as they seal. It's like music to a homesteader's ears!

8. After the twenty-four-hour period, you can remove the bands, wipe any residue off the jars, and line them up on your shelf for storage. Go, you!

It would require an entire book to cover everything my family cans during a given season, but I'm sharing a few of my favorites with you. When you enjoy them, it'll be like I'm there with you (in the least creepy way possible, of course).

The Best Pickled Asparagus

The title of this recipe may lead you to believe that this is the best pickled asparagus recipe. And that, my friends, would be a correct statement.

For each quart jar, you will need:

> **1 clove of garlic**
>
> **½ teaspoon organic red pepper flakes**
>
> **¼ teaspoon sea salt**
>
> **1 teaspoon organic dill seeds**
>
> **1 teaspoon organic dried oregano**
>
> **1 teaspoon organic mustard seeds**
>
> **½ teaspoon organic black peppercorns**
>
> **10 to 20 spears of fresh and local asparagus, depending on spear size**

1. Wash and sterilize the mason jars. Line them up on the counter like little ducks in a row.

2. Into each jar, put the garlic, red pepper flakes, sea salt, dill seeds, dried oregano, mustard seeds, and black peppercorns.

3. Wash the asparagus and cut the bottom of the stem off so that the asparagus fits into the quart jar standing up. Gently stuff each quart jar with as many asparagus spears as it will hold. You should now have quart jars full of spices and asparagus spears, correct? Correct.

4. Heat up a vinegar brine of 50 percent filtered water and 50 percent white vinegar to near boiling. The amount you need will vary depending on how many quart jars you are processing.

5. Gently pour the hot vinegar brine into each asparagus-filled quart jar, leaving ¼ inch headspace at the top. Secure a sterilized lid and band onto each quart jar.

6. Process the asparagus for 15 minutes in a boiling-water-bath canner. Gently remove and set on the counter to seal.

Roasted Tomato Salsa

In the middle of winter, you'll no doubt be thankful you decided to take a few hours to make this homemade salsa. It's the perfect way to use up that surplus of tomatoes you've spent all summer growing!

10 cups roasted tomato puree

2 cups finely chopped onion

2 cups finely chopped sweet bell pepper

⅔ cup finely chopped mild chiles

2 tablespoons chopped cilantro

2 tablespoons minced garlic

1 cup apple cider vinegar

2 teaspoons sea salt

4 teaspoons ground cumin

2½ teaspoons dried oregano

1. **To make the roasted tomato puree:** Wash the tomatoes and cut in half. Place cut side down on a baking sheet. Roast in a 425-degree F oven for 30 to 45 minutes or until they release liquid. Carefully remove the baking sheets from the oven and pour out some of the water. Put back into oven and continue to roast until the tomatoes are just slightly browning on top, about 15 more minutes. Remove from oven and let cool to room temperature.

2. Once the roasted tomatoes have cooled, put them into your food processor or blender. Pulse until they're your desired texture (I like a smoother salsa so I tend to blend mine quite a bit. If you like a chunkier salsa, simply blend them less). Pour the pureed tomatoes into a large bowl.

3. In a large pot, combine the tomatoes, onions, peppers, chiles, cilantro, garlic, vinegar, salt, cumin, and oregano. Bring to a simmer and continue to simmer for 10 minutes.

4. Carefully ladle the hot salsa into sterilized, warm glass canning jars. Add a lid and band to each jar. Process for 15 minutes in a water canner.

5. Remove and let cool at room temperature until the lids have sealed.

Honey Peaches

Yes, peaches can be canned in honey. Lawd have mercy, are they ever delicious. We avoid white sugar like the plague around these parts, so canning our fruit in honey is a wonderful alternative.

> **Roughly 3 pounds of peaches per quart jar**
>
> **Honey syrup (recipe below)**

1. Peel the peaches. To make this easier, simply dunk the peaches into boiling water for a minute until the skins easily peel off. Pit the peaches by cutting them in half and removing the pit. You may cut the halves into pieces, if desired, or leave the halves intact.
2. Place the peaches into sterilized, warm, glass canning quart jars. Pour honey syrup over the top, leaving ½ inch headspace.
3. Add a lid and band to each jar.
4. Process in a water canner for 25 minutes. Remove and allow the jars to sit on the counter overnight to cool and seal before transferring to storage.

Honey Syrup: In a large saucepan, combine water and honey at an 8 to 1 ratio (8 cups of water to every 1 cup of honey). Cook over medium heat until combined.

Honey makes a wonderful alternative sweetener for your canned goods.

Vanilla-Infused Cherries

The first time I ever ate a canned cherry, I was in disbelief at how delicious it was. How had I lived my entire life without devouring at least a jar a day? In the doldrums of winter, there's hardly anything more welcomed on my pancakes.

Roughly 2 pounds of cherries per quart jar

Whole vanilla beans

Honey syrup (see recipe above)

1. Pit the cherries, if desired. Sometimes I'm feeling awesome and pit them all. Sometimes, I just throw them into the jars as is and deal with the pits while I'm eating or cooking with them later on. Up to you! A $20 pitter will be worth its weight in gold, though, should you decide to pit them all.

2. To the bottom of a sterilized, warm, glass canning jar, add a quarter of a vanilla bean. Then, stuff the rest of the jar with cherries!

3. Top off each quart jar with honey syrup, leaving ½ inch headspace.

4. Add a lid and band to each jar.

5. Process in a water canner for 25 minutes. Remove and allow the jars to sit on the counter overnight to cool and seal before transferring to storage.

Pressure Canning

A great option for preserving low-acid foods (with a pH higher than 4.6), including meat, is to use a pressure canner (also known as a pressure cooker). The acidity of a food is best determined by following your pressure canner's instruction guide or a canning reference manual. Like the name denotes, pressure canning uses pressure to bring the temperature of the food past boiling point, up to 240 degrees F. This is high enough to kill even the most stubborn bacteria, including *Clostridium botulinum* (commonly known as botulism). Pressure canning can often be a bit nerve-wracking to the novice, who might imagine the pressure canner lid somehow affixed to her kitchen ceiling. Luckily, pressure canners are now made with safety valves and gauges that make using them a breeze. Follow the directions, watch your pressure gauge, and you'll be good to go. I see beautiful canned corn in your future . . .

Jars of produce can be safely canned using just a pressure cooker!

The Basics

Like a water canner, a pressure canner is a large pot with a lid, often made from stainless steel or aluminum. Pressure canners have been designed with a pressure gauge and seal specifically for maintaining high pressure within the canner. Food is packed into sterilized jars with metal lids and bands. The canner is filled with a small amount of water, and the jars are placed atop a rack in the canner. After the lid is put on, the heat is turned up, and a particular pressure per square inch (PSI) is achieved by capturing the steam within the canner. Like water canning, the jars are then processed for a given amount of time, before the pressure is gently released and the jars are removed.

The Catch

Pressure canning requires a bit of focus on your part. It's important that a particular pressure be achieved for various foodstuffs, so concentrate on your recipe and know what pressure you're working toward. The pressure is regulated by the heat source. More heat will create more steam and thus, more pressure. Once the proper pressure is achieved, it's easy to maintain by manipulating the heat source accordingly.

Which Method to Choose?

Water canning works best for high-acid foods pH 4.6 or higher such as fruits, jams, jellies, pickles, tomatoes, salsa, or syrups.

Pressure canning works best for low-acid foods (pH 4.6 or lower) such as meat, soups, chili, vegetables, broth, fish, or sauces.

The Method

Pressure canners are slightly more expensive than water canners, but since they can actually double as water canners (just don't seal the lid all the way!), they're still a great investment for your homestead. Vegetables, fruits, legumes, meats—imagine the possibilities! Here are the supplies you'll need:

Pressure canner

Glass jars

Metal lids and bands

Rack

Jar tongs

Tea towel

1. Pack the foodstuff, according to your recipe, in sterilized glass jars.

2. Wipe off the rim of the jar to ensure a good seal.

3. Place the metal lid on the jar, seal side down, and twist on the metal band.

4. Place the jars into the pressure canner and add 2 to 3 inches of water. Secure the lid on the pressure canner. Turn the heat up underneath the pressure canner until steam begins to escape from the valve. Allow a steady stream of steam to escape for 10 minutes, then put the weight over the steam valve to contain the steam within the canner. The pressure gauge will begin to rise at this point. Watch it carefully to achieve the correct PSI for your recipe. Adjust the heat as necessary to maintain the correct PSI for the amount of time your recipe states.

5. Once the set amount of time has passed, turn off the heat, and allow the canner to sit undisturbed until the pressure gauge has dropped down to "normal" or "zero." This usually takes between 30 and 45 minutes.

6. Carefully, with a towel or pot holder, remove the weight from the steam valve. Allow the canner to vent any leftover steam for 10 to 15 minutes before removing the lid.

7. Carefully remove the jars from the canner utilizing jar tongs, and set them on a tea towel to continue cooling. See! I told you you could do it! You're a rock star.

We use our pressure cooker for a variety of dishes and preserves. Here are a few of the basic recipes that I started with to help to get you started in using yours!

Basic Green Beans

Ah, pressure-canned green beans—just the goodness of fresh beans, filtered water, and a bit of sea salt. My kids devour them like candy.

Fresh green beans

Sea salt

Filtered water

1. Wash the green beans. Sit down with a glass of chai tea and your favorite radio station and get to work cutting off the teeny tops and bottoms of the beans (I hate biting into one of those little things on the end. They gag me!). Some people may choose to skip this step, but I never do. Because I don't like gagging.

2. Break the beans into 1- to 2-inch pieces, using either your fingers or a knife.

3. Gently pack the cut green-bean pieces into jars. Either quart jars or pint jars will do just fine.

4. After you've filled the jars with green beans, add ½ teaspoon sea salt to each pint jar OR 1 full teaspoon sea salt to each quart jar. Don't skip the salt. Only blandness will come of that. And that's not how we roll.

5. Boil a large kettle of filtered water. Once it's boiling, pour over the top of the green beans, leaving 1 inch of headspace in each jar. Wipe the rim of the jar to remove anything that could prevent it from sealing. Add a new lid and band onto each jar and gently tighten.

6. Place the jars on the rack in a pressure canner and add the designated amount of water for your pressure-canner model. Seal the lid on the pressure canner, place it on the stove, and turn up the heat to medium-high.

7. Process at 10 pounds of pressure for 25 minutes.

8. Carefully remove the jars. Set them on a baking sheet or towel and allow them to cool for 12 to 24 hours before transferring to long-term storage.

If the pressure ever falls below the recommended PSI during the processing time, turn the heat up to obtain the correct PSI, and begin timing again from the very beginning. This will ensure the food is safe!

Chicken Stock

Stock is a product that goes rancid quickly in the refrigerator, so canning is a wonderful option to always have it available and ready to use!

Chicken carcass (including feet, head, bones, etc.)

Vegetable scraps (onion, carrots, celery, etc.)

Peppercorns

Apple cider vinegar

1. In a large stockpot, combine the chicken carcass, vegetable scraps, a pinch of peppercorns, and a generous glug of apple cider vinegar. Cover the contents completely with filtered water. Bring the stockpot to a boil and allow the stock to simmer for 12 hours.

2. While the stock is still hot, strain it into clean, glass jars leaving 1 inch headspace.

3. Wipe the rim of the jar to remove anything that could prevent it from sealing. Add a new lid and band onto each jar and gently tighten.

4. Place the jars on the rack in a pressure canner and add the designated amount of water for your pressure-canner model. Seal the lid on the pressure canner, place it on the stove, and turn up the heat to medium-high.

5. Process at 10 pounds pressure for 25 minutes.

6. Carefully remove the jars. Set them on a baking sheet or towel and allow them to cool for 12 to 24 hours before transferring to long-term storage.

Home-Canned Beans

Now, I know store-canned beans are crazy convenient. But they're also a zillion times more costly than dried beans and contain excess amounts of salt and in some cases, even preservatives! So instead of splurging, I always buy my beans in bulk from a local farm and can them myself.

Dried organic beans of your choice

Filtered water

1. Pour the desired amount of dried beans into a large bowl.

2. Cover the beans with a large amount of water. They will soak up *a lot* and in order to get them hydrated, you need to provide them with plenty of aqua.

3. Set the bowl of beans out on your counter and ignore them for the next 12 to 24 hours. Now, isn't it nice to ignore things once in a while?

4. After the beans have hydrated, strain out the extra water and pour the beans into a large stockpot. Cover with fresh water and bring the beans to a boil to get them nice and hot.

5. After they've reached a boil, carefully (carefully now!) scoop the hot beans into pint or quart jars.

6. Ladle the hot water over the top of the beans, leaving 1 inch of headspace in each jar. Wipe the rim of the jar to remove anything that could prevent it from sealing. Add a new lid and band onto each jar and gently tighten.

7. Place the jars on the rack in a pressure canner and add the designated amount of water for your pressure-canner model. Seal the lid on the pressure canner, place it on the stove, and turn up the heat to medium-high.

8. Process in the pressure canner at 10 pounds of pressure for 1½ hours.

9. Carefully remove the jars. Set them on a baking sheet or towel and allow them to cool for 12 to 24 hours before transferring to long-term storage.

Berries lend themselves very well to freezing and are delicious added to winter oatmeal, cakes, or smoothies.

Freezing the Harvest

I must admit, I'm quite fond of freezing the surplus harvest. Partly because I'm a busy woman and it takes less time and effort than other methods, and partly because I appreciate the ease of grabbing "fast food" from my downstairs freezer. When I don't have the time for water or pressure canning, or when I've plum run out of energy at the end of the season, freezing the harvest is where I turn.

Frozen fruits and vegetables maintain their flavor, color, and nutrient content well—particularly berries, greens, herbs, onions, and even potatoes! Freezing requires very minimal equipment: containers, freezer bags, a knife, and perhaps a large, well-loved pot for a bit of blanching are all that is needed.

Here are a few of the staples that I make sure to put up each season. These frozen goods keep us stocked and eating well throughout the winter and take almost no time at all!

Frozen Berries

Because there's nothing better in your winter oatmeal than frozen berries. *Nothing.*

1. Wash the berries and remove any debris.
2. Lay out the berries in a single layer on a baking sheet lined with parchment paper.
3. Place the baking sheet in the freezer for 2 hours, until the berries are just frozen (this will keep them from sticking together!).
4. Transfer the berries to a plastic bag or storage container of your choice and keep in the freezer until needed.

Diced Onions and Peppers

One of the best ways to add flavor to dishes is with onions and peppers. This makes them an easy addition to a variety of dishes no matter what time of year it is!

1. Peel the onions, deseed the peppers, and dice them all up into small pieces.
2. Lay out the onions and peppers in a single layer on a baking sheet lined with parchment paper.
3. Place the baking sheet in the freezer for 1 hour, until the onions and peppers are just frozen.
4. Transfer to a plastic bag or storage container of choice and keep in the freezer until needed.

To save even more time in the kitchen, consider doing a bit of cooking before putting that produce in the freezer! For example, sautéing your onions or roasting your garlic pre-freezing can be a great way to save time later on.

Olive-Oil Herb Drops

If you've got fresh herbs, this is one of the wonderful ways to preserve them (check out a few other great methods at the end of the chapter). Frozen herbs still aren't quite as awesome as fresh ones, but they're fantastic to have on hand when fresh herbs aren't available!

Herbs of choice

Olive oil

1. Wash and mince the herbs.
2. Place the herbs into an ice cube tray. Stuff that tray full, baby!
3. Fill up each ice cube container with olive oil (you may need to push the herbs down a bit to keep them submerged).
4. Freeze until set before popping the cubes out of the tray. Store the olive-oil herb drops in a freezer safe container. Grab a cube and throw it in the skillet whenever you need one!

Shredded Potatoes

Yes, potatoes can be frozen. How cool is that? If you don't have a cool place to store your extra crop, freezing potatoes can be a great way to keep 'em delicious.

1. Peel the potatoes, if desired. Then, use a box cheese grater or food processor to shred the potatoes.
2. Blanch the potatoes in a large pot of water for 3 to 5 minutes. Remove with a slotted spoon and put them into an ice bath until cool.
3. Scoop the potatoes into a large colander. Use your hand or the back of a wooden spoon to smooth the potatoes. This will drain the excess liquid out of them.
4. Line a baking sheet with parchment paper. Pile the potatoes into hamburger-patty-sized stacks, pressing them down a bit to create a disc shape. This makes it super easy to grab a portion of potatoes out of the freezer without having to defrost them all! Stick the baking sheet into the freezer for a few hours until the potatoes set.
5. Remove the potato stacks from the sheet and pop 'em into a freezer bag.

Apple-Pie Filling

I'm from the apple capital of the world! What can I say? We eat a lot of apples. Freezing pie filling is an annual undertaking around these parts, and though it takes a bit of prep to get those bags in the freezer, pie is always a welcome treat in the middle of winter when we're all hungry for something fresh and comforting!

6 tart apples, peeled, cored, and sliced

Juice and zest of 1 lemon

⅔ cup honey

¼ cup maple syrup

1½ teaspoons cinnamon

¼ cup flour

½ teaspoon sea salt

1. Combine the apple slices, lemon juice, and lemon zest together in a saucepan. Stir to coat the apples. Then, add in the honey, maple syrup, cinnamon, flour, and sea salt. Heat over medium heat and stir continuously until the honey melts and the mixture is well combined. Taste and adjust the seasonings as needed (apples are all different!).

2. Line a pie pan with parchment paper. Fill the pie pan with the apple pie filling and set it in the freezer for a few hours to set. This freezes the apples into the shape of the pie, so when it comes time to bake the pie, it's easy to slip right in! Remove the filling from the pie pan and transfer into a plastic bag. Place in the freezer until needed.

Pesto is a wonderful way to capture the flavor of summer basil!

Freeze according to usage portion. If you freeze your pesto in a gallon-sized jug, it'll be a pain to haul it out and thaw a bit each time you'd like to smear some on a fresh pizza crust! Instead, freeze your goods in the portion that you'll most commonly use.

Pesto

The stuff midwinter dreams are made of. The delicious sauce that reminds you during the cold, dreary months that soon there will be life from the land once again. For pizzas. For pastas. For omelets. For smearing on homemade bread. And for all the delicious things in between.

2 cloves garlic

4 cups tightly packed fresh basil leaves (no stem or flowers!)

¼ cup cubed Parmesan cheese

1 cup olive oil

½ cup almonds, pine nuts, or walnuts

Sea salt to taste

1. In a small skillet over medium heat, gently toast the garlic cloves (with their skins still on) until just golden and fragrant. When cool, remove skins.

2. In a high-powered blender or food processor, combine the garlic cloves, basil, Parmesan, olive oil, and almonds. Blend on high until very smooth, scraping down the sides of the blender or food processor if necessary. Personally, I don't love chunky pesto, so I make sure to blend it really well.

3. Taste the pesto and season with a small pinch of sea salt, if necessary.

4. Well, I guess that's it. Not too bad, huh? Freeze in small containers to use throughout the winter or keep it in your refrigerator for immediate use.

Freezer Cherry Jam

Cherries are in abundance around these parts, and this quick and easy freezer jam is a great way to capture their deliciousness for the coming months.

18 cups washed, pitted, roughly pureed sweet cherries (or berries of choice)

7 cups organic, whole dehydrated cane sugar

3 cups filtered water

10 teaspoons pectin

1. Add all the pureed cherries into a large bowl.
2. Add in the sugar. Stir to combine. Dip your clean finger in and lick it off. Sweet enough? Perfect.
3. In a small saucepan, heat the water to a boil. Add in the pectin and use an immersion blender (or transfer to a regular blender) to combine. The pectin will become quite sticky and will clump together, so it's really handy to have a blender on hand to whip it together as quickly as possible. Once the pectin is dissolved, remove from heat.
4. Pour the pectin/water mixture into the cherry mixture.
5. You'll see the jam begin to set almost immediately. Just keep stirring for a few minutes while it all combines.
6. Scoop out the jam into freezer-safe containers.
7. Allow the containers to sit and settle at room temperature for a few hours. After the jam has completely set, transfer to the freezer.

Plan to use up all your frozen goods within 8 to 12 months for the best flavor.

Dehydrating the Harvest

A very common and popular method of preservation is dehydration. Quite simply, dehydrating involves removing water from the produce. What's left is a shelf-stable product that you can enjoy year-round. Dehydrating requires a bit more work, and a bit more time, than canning or freezing, but results in a product that can be stored at room temperature. This is a big perk if extra freezer or storage space is an issue for your homestead. There are a variety of dehydration methods that make it achievable for anyone, even without any special equipment. Dehydrated fruit, in particular, is a welcome treat for our kiddos. I bring out "fruit candy," as they like to call our dehydrated pears, on special occasions and it is a wonderful reflection of the summer and fall bounty.

Commercial Dehydrators

The most common dehydration method used these days involves a commercial dehydrator. These are machines specifically designed to dehydrate your harvest easily and efficiently, be it fruit, meat, vegetables, or herbs. Dehydrators come in a range of sizes with a large variety of settings and can be as fancy as you'd wish. My first dehydrator was a third-hand one passed down to me from my aunt (she'd picked it up at a garage sale decades before). Only a few of the settings work, and there is a mangled piece of duct tape keeping the door shut, but it's still my go-to machine for dehydrating. Dehydrators can often be picked up secondhand for a fraction of the cost of a new one. That being said, new dehydrators come equipped with large trays, efficient circulation, timers, and specific temperature control–all great traits to have in a dehydrator! One of the biggest perks of a commercial dehydrator, often stocked with ten or more dehydrating trays, is the amount of produce it can dehydrate at any one time. My dehydrator continually lives on my counter from summer to fall, gently humming as it works, and sends out sweet aromas and a bit of warmth from its vents. My love for it knows no bounds.

Some produce will remain pliable, such as tomatoes or apricots, while other produce, such as kale or peas, will become extremely dry and hard. Have an idea of what you're trying to obtain before you get started so you know what to watch for. As a general rule, fruits will remain soft and vegetables will become brittle.

Basic Method

Prepare the fruit or vegetables by slicing them $^1/_4$ inch thick, discarding any old, squishy, or moldy parts. Lay the slices spaced out on the dehydrator trays, allowing enough room for air to circulate completely around each piece. Set the temperature according to the dehydrator specifications and run until the produce is done. This can take anywhere from six to twenty-four hours. As you gain experience in utilizing your dehydrator, you'll soon be able to recognize the correct "doneness" of the produce, simply by touch.

Sun Drying

The oldest method of dehydration involves "solar power"–that is, the sun! Traditionally, summer involved spreading fruit on old bed sheets or drying racks in the sunshine so that the warmth from the sun would naturally dry out the goods. Meats and vegetables were not typically dehydrated in this way, as they are much more susceptible to spoilage than their high-acid fruit friends. Humidity also plays a big role in utilizing the sun for dehydration. Ideally, humidity should be below 60 percent, as excess humidity in the air encourages spoilage. This can make sun-drying tricky in the southern United States. Much of the fruit available for preservation also tends to show up in the fall (hello, apples!), which is not exactly an ideal time of year to capture all those super-powerful sun rays. That being said, if you've got a surplus of summer fruit and the ideal weather conditions, sun-drying can be a great way to dehydrate the harvest.

Basic Method

Prepare the fruit or vegetables by slicing them $^1/_4$ inch thick, discarding any old, squishy, or moldy parts. Lay them spaced out on an old sheet or drying rack, allowing enough room for air to circulate completely around each piece. Place in a sunny, hot, and safe location (by safe, I mean somewhere a neighborhood dog or wandering child can't easily disrupt). Allow the produce to remain outside for several days (over 90 degrees F during the daytime is ideal) until it's completely dehydrated and ready for storage.

Cover sun-drying produce with cheesecloth (or a second drying screen) to keep out birds! They'd love a tasty nibble of your summer strawberries.

Conventional Oven

We have the great privilege in our day and age of being able to use conventional ovens for dehydration. With the push of a button, we can control temperature. To think what homesteaders of an earlier era would've given for such a convenience! Most conventional ovens will not go lower than 170 degrees F, which is still a great temperature for dehydrating certain foodstuffs. I primarily use our oven to dehydrate soaked and sprouted nuts, but I've also used it for a variety of fruits when the dehydrator was full and the baskets were overflowing! The biggest downside to dehydrating in an oven is the amount of space available. It's hard to work through a mountain of produce when there are only a few racks in the oven.

Basic Method
Prepare the fruit or vegetables by slicing them ¼ inch thick, discarding any old, squishy, or moldy parts. Lay them spaced out on the wire racks, allowing enough room for air to circulate completely around each piece. Set the temperature as low as the oven will allow. Place the produce in the oven, crack it open a few inches, and allow the produce to dehydrate until it's done. This can take anywhere from six to twenty-four hours.

Cold Storage of the Harvest

One of my favorite ways to store excess produce is through cold storage. Just like the name suggests, this method utilizes cool temperatures to preserve food–the same concept as a refrigerator. Cooler temperatures encourage the food to break down (i.e., rot) at a much slower pace, thus extending its life. I'll be honest with you here. The reason this method rocks my socks off is because it requires almost no effort on my part. Instead, one simply needs to rely on the coolness of a storage shelter, root cellar, refrigerator, or even a mound of soil in the garden to do the work. Even for the small-time backyard gardener, this can be a great way to carry over a surplus of root vegetables with very little planning or prep work.

If it starts to rain on your sun-drying party, bring the goods inside and finish them in the oven!

Basic Root-Cellar Storage

Naturally, there are particular crops that lend themselves better to cold storage than others. Potatoes, onions, shallots, beets, garlic, squash, carrots, cabbages, parsnips, and turnips are at the top of the list! By their nature, root vegetables are essentially large energy vessels for the plant. This means that if harvested at the proper time and stored well, we can preserve that energy! Even among these crops, there are particular varieties that tend to store very well, so select specific varieties that lend themselves to long-term storage if this is a route you'd like to explore. For example, I grow a variety called Music garlic almost exclusively because of its storage qualities. Each summer, I pull all the garlic heads from the ground, allow them a few days to "cure" in the sunshine, before tying the heads up into braids and hanging them in the cellar–leaving a few by the kitchen stove for easy access, of course. We are able to enjoy our homegrown garlic well into the winter this way! Apples are one of my favorite crops to store in the root cellar because we have such easy access to them, and frankly, it's much easier to throw a bin of apples in the cellar than it is to can all that fruit! That's not lazy homesteading–that's smart homesteading!

I remember being young and venturing down to my grandpa's root cellar. It sat in the basement of his home, built sometime in the 1940s when preservation was still practiced, and was lined with shelves of produce, flower bulbs, and canned goods. There were always heavy cobwebs hanging from the ceiling and the smell of moist earth in the air. I hated going down there because–hello!–cobwebs and dark, dank basement cellars are super creepy to a six-year-old. Now that we've got a root cellar ourselves, I can't wait to creep my children out with the same memories. (I'm such a good mom.)

The principles can vary ever so slightly from crop to crop, but for the sake of simplicity, let's focus on the basics:

Some crops, such as onions and garlic, prefer cool and dry storage. Don't store these in your humid root cellar!

Always lift the bins of produce a few inches off the ground so that the air can easily circulate around all of the produce. This helps to prevent mold.

> *Keep it cool.* The idea of root-cellar storage is to store the produce in a cool environment, much like a refrigerator. Back before refrigerators existed, this was the way people kept much of their produce from decomposing too quickly. Ideally, the produce will be held at an average of 52 degrees. It's great to understand how air moves when designing your root cellar: warm air rises and cool air drops. To keep the area at the correct temperature and ventilated, it's ideal to have a large outlet for the warm air and a small outlet to draw in new, cool air.

> *Keep it moist.* If the air in the root cellar is too dry, the air will draw moisture from the produce and will result in shriveled produce all too quickly. The way to prevent this is to keep the humidity at around 90 percent. Yes, that's pretty high! But a true root cellar will have a dirt floor, which will naturally hold moisture, keeping the humidity high. Gently wetting the floor can be a great way to increase the humidity. You can also place damp towels over the top of the produce bins to maintain good moisture in the air.

> *Keep it dark.* Produce stores much better out of direct sunlight, which will break it down quickly. Sunlight also encourages the roots to send up shoots, which renders the food inedible! Keep the root cellar free of windows!

DIY Root Storage

Our little cottage was built in 1909. There are stairs that wind down into a room under the kitchen that is a two-room root cellar. It's cool in the summer, insulated by the dirt that surrounds the walls. It's also cool in the winter but warm enough to keep the produce from freezing. If you don't have a root cellar in your house, which most houses built after modern refrigeration don't, have no fear! Here's a cheap and easy method to mimic this same idea on your own patch of heaven:

If you need even more moisture in the air, consider storing the crops in bins of moist sawdust, leaves, sand, or straw.

Garbage Can Root Storage

Just like it sounds, this method utilizes regular ol' metal garbage cans to make a root storage area right in your own yard. Metal is great for keeping rats and mice from chewing their way in.

1. Pick a location in your yard. Ideally, your root storage garbage can shouldn't be more than ten or twenty feet from your kitchen. This will make it easy to grab what you need, even when winter is at its worst!

2. Dig a gigantic hole (big enough for a garbage can).

3. Drill a dozen or so holes into the bottom of a clean trash can to allow excess water to drain from the can.

4. Bury the garbage can, leaving a 4-inch rim above the surface of the soil. Use dirt to fill in around the hole so that the garbage can is packed tight into the soil.

5. Place a generous layer of straw, wood chips, or leaves at the bottom of the garbage can, then a layer of produce. Add another layer of straw, followed by another layer of produce. Continue layering until the garbage can is full, and top off with a thick layer of straw. Put the lid on the can and secure it by placing a few bricks on top to keep wandering animals from getting in. A generous layer of mulch over the top will add extra insulation and should be easy enough to brush aside when you venture out in February to grab a week's worth of cabbage (plus it makes it look prettier, which is certainly important!).

If you're storing a few varieties of the same vegetable or fruit, put the longest-lasting variety on the bottom of the bin for best results.

Common Cold-Storage Crops

Naturally, some produce stores better than others. There are a few vegetables and fruits that are ideal candidates for cold storage.

To figure out how much of each crop you'll need to store, calculate how many pounds of each crop your family consumes in a week and how many weeks your average winter and spring last. Proper planning can keep

How to Braid Onions

I like to braid my onions for a variety of reasons. First, it's efficient because the weight of the onions is distributed evenly versus having a few heavy onions on top squashing all the onions underneath. No onion victims are squished in a braid. On top of that, it's aesthetically pleasing to the eye (at least for this homesteader), and it's a great way to get the onions out of cabinets or bins, and up off the ground where they have a better chance of not being eaten by rodents or used as bowling balls by one of your little rascals. The same process can be replicated with garlic, as well.

Storage onions

Twine

Scissors

1. Pull the onions from the garden bed and allow them to sit in the sun for a day or two to allow the onion's skin to develop on the outside of the onion. This will protect the onion during storage. Don't skip this step!

2. Line up three onions. Cross their tails, just like you're going to begin a braid.

3. Gently fold the left onion's tail toward the middle. Repeat with the right onion's tail.

4. Add another onion to the middle. Braid the left tail and then the right tail toward the middle once again.

5. Repeat, adding onions as necessary, until the braid is holding 10 to 12 onions (depending on size and weight). This isn't an exact science. Rather, it's an art form.

6. Once the braid is holding a dozen or so onions, use a long piece of twine to tie it off *tightly*. The tails of the onions will continue to dry and shrink, so it's important to ensure a snug knot.

7. Hang in a cool, dry location, cutting off an onion as needed (but leave the green tail in the braid!). Always use damaged onions first as they will store for a much shorter period of time.

Leave the onion tops on to make them easy to braid and store through the winter months.

a root cellar stocked all winter long! For a family of five, here are some averages to keep you with fresh produce throughout the winter:

Apples: 100 pounds

Pears: 30 pounds

Cabbage/Chinese cabbage: 30 heads

Carrots: 75 pounds

Leeks: 5 pounds

Potatoes: 125 pounds

Turnips: 20 pounds

Parsnips: 20 pounds

Beets: 30 pounds

Onions: 50 pounds

I always keep a small basket of storage onions in the kitchen for easy access!

Culturing the Harvest

I'm a traditional foodie, which means that I've developed a special love for foods that have been naturally cultured to perfection. Culturing is an age-old method of food preservation that was used well before the invention of canners and freezers. Culturing ferments the product with the help of beneficial bacteria, bringing it to a new level of awesomeness. Cultured foods contribute to our gut health and aid in the digestion and absorption of nutrients through their bacteria, enzymes, B vitamins, and probiotics! Each year, you'll find us culturing garlic, carrots, beets, tea, garlic scapes, berries, and most importantly (at least in this house), sauerkraut and kimchi! It's the perfect way to preserve summer cabbages. The same method can be performed for any produce you've got coming out of the garden.

Homemade Fermented Sauerkraut

1 or 2 medium heads of organic cabbage

2 tablespoons sea salt

3 cloves garlic, minced

½ teaspoon red pepper flakes

1. Using a knife or a food processor, shred the cabbage into thin strips. Place in a large bowl.

2. Add the salt, garlic, and red pepper flakes. The salt will begin to draw the liquid from the cabbage. It's magic!

3. Use a wooden spoon to smush the liquid from the cabbage. This takes about ten minutes. Get it nice and pulverized and juicy.

4. Wash and dry a half-gallon glass jar. Then, place a bit of the cabbage into the jar. Use the wooden spoon to help press the cabbage tightly into the jar. Add more cabbage, then squish some more. Add more cabbage, than squish some more. You get the idea? After all the squishing, the liquid should reach the top of the cabbage. If it doesn't, press down some more so that all of the cabbage is submerged in the liquid. The liquid should remain at least 1 inch *below* the top of the jar so there is room for some expansion during the fermentation process. Cover the jar with a lid.

5. Let the jar sit out on your counter for three days, at which point, it can be moved into the refrigerator. Like most fermented vegetables and fruits, the flavor will increase and intensify over time.

Drying and Storing Herbs

Lucky for us, herbs are easy as pie to dry and store for later use. This is great news for those of us that spend November through March under powdery snow and dark skies. Most herbs peak in summer, so plan on drying and storing during the later part of the summer. Properly dehydrated, herbs can stay delicious for months and months. Dried herbs are stronger in fragrance than their fresh counterparts, but the fragrance will gradually dissipate the longer they are stored, So plan on using them up the winter and spring after they're stored.

Basic Drying Techniques

1. *Harvest your herbs.* If you prefer to dry your herbs in bunches, you can harvest them by clipping them off at ground level and keeping the leaves on the stems. Alternatively, you can carefully pull off the leaves and gather them in a basket.

2. *Dehydrate the herbs.* The various methods listed below can accomplish this, but the purpose is the same: remove the moisture content from the herbs. Moisture equals spoilage. And we don't want any of that, do we?

    Air drying. "Dry herbs," such as sage, thyme, lavender, bay, lemon balm, and mint, can easily be air dried by simply bundling the herbs together and hanging them in a well-ventilated, dry area.

    Dehydrator. If you live in a humid area, the dehydrator can be a great way to remove the moisture. Dehydrate herbs on a low setting until the herbs are crispy to the touch, rotating trays part of the way through if needed to ensure even drying.

    Oven drying. Lay the herbs in a thin layer on a baking sheet and place them in an oven on the lowest setting. High heat will damage the oils in the herbs, so keep it low and slow. Most herbs take about an hour to dry. Turn them halfway through to ensure they're drying evenly.

3. *Store the herbs.* Use your fingers to crumble the herbs into small pieces, removing any bits of stem, and transfer to an airtight container for storage. I do this over a large bowl as it can get a bit messy with all those little pieces! Herbs will lose their color if exposed to sunlight, so it's a good idea to keep them in a dark area or, alternatively, in dark airtight containers.

Tea Time

I love homemade herbal teas. I mean, come on, you can grow your own tea, y'all. If you do, your coolness factor will go up in the world.

To enjoy these teas, simply boil the water and steep the herbs in the water for 3 to 5 minutes. Sweeten with honey, if desired, sip, and savor!

Enjoy homemade tea? Plant these herbs and plants in your garden:

Catnip

Chamomile

Dandelion

Echinacea

Lemon balm

Lemongrass

Lemon verbena

Orange mint, pineapple mint, chocolate mint, spearmint, and peppermint—oh my!

Rosemary

Raspberry leaves

Rose hips

Peppermint Tea

1 teaspoon dried peppermint leaves

1 cup filtered water

Lavender-Lemon Tea

1 teaspoon dried lemon balm

½ teaspoon dried lavender flowers

1 cup filtered water

Chamomile Tea

1 tablespoon chamomile flowers

1 cup filtered water

A Note on Farm Kids

I remember the first moment I realized that I'd raised a farm kid—the irony being, of course, that we didn't live on a farm at the time. Back in Alabama, on our little patch of sand, we kept a hutch of rabbits in the backyard. Each morning, after Stuart left for work, our three-year-old daughter Georgia and I would grab our buckets and head out on a walk around the neighborhood to gather weeds for the rabbits. We'd spend a few minutes feeding the rabbits. Georgia liked to talk to Scooter, her favorite. When it came time to butcher our first pair of rabbits, she calmly watched these animals that she'd spent time and energy growing become our dinner. Later that same week, a bird hit our window and died. "Are we gunna eat it, Mama?" she asked me.

I was proud. Even at a young age, she had demonstrated the understanding that an animal's life is valuable and not to be wasted. Even though we spent time caring for our rabbits, it was all with a purpose. Her little mind was beginning to comprehend what that meant.

Our four farm kids now run around like wild animals most days, though we are still able to rope them into plenty of farm chores. Learning to farm with kids is part of the fun, and let's be honest, part of the frustration. It can become dangerous to plow forward at maximum speed without taking the time to have them alongside you, but remember, this is formative for them. This is going to shape how they see the world, how they see food, and how they learn to work hard. In past years, we've welcomed a variety of my husband's students up to our farm to work. Seeing their faces and hearing their questions never ceases to amaze me! And it reminds me that everyone needs to understand that male animals can't, won't, and will never be able to make milk.

Here's what the little ones can do around the farm:

▷ *Collect eggs.* Yes, they'll break some. But they'll soon learn the value of an egg!

▷ *Fill waterers with the hose.* It's a great first step in learning to care for the animals.

▷ *Participate in harvesting.* In my experience, kids can be great partici-
pants on butchering day. They begin to learn about death, harvest,
appreciation, and culinary skills at their very basic level.

▷ *Check feed.* They may not be big enough to throw the livestock hay, but
they can surely tell you if the rabbits need more feed or if the layers
are out of grain. Make it their job to report it.

▷ *Help feed hay.* If they're big enough, put 'em to work with feeding hay,
even if that means a few flakes at a time in a wheelbarrow or wagon.

▷ *Clean pens.* It may take a while before they're actually any help, but
that shouldn't keep you from handing them a rake and shoving them
in the chicken coop.

▷ *Milk the dairy animals.* I dream of the day my kids are old enough
to learn this valuable skill! Kids tend to cut corners, though, so I'd
always recommend supervising them until you're positive that all
milking protocol is being correctly followed.

▷ *Weed the garden.* I wouldn't recommend this in the spring when all
those teeny tiny seedlings are coming up, but when the plants are
established and the child can easily distinguish between the rows of
produce and weeds, they're more than capable of helping with this
task. Sometimes I bribe mine: "Fill the bucket with weeds for a piece
of chocolate?" Heck, I'd do it.

▷ *Help you preserve food.* My oldest, who is only six, is a great help at
pitting cherries and apricots, slicing cherry tomatoes, and smashing
sauerkraut. It always leads to fun conversations about how we'll be
enjoying the food later on that year!

▷ *Plant their own garden.* Maybe it's the control freak in me, but I just
can't get myself to let the kids help me plant *my* garden. I'll happily let
them plant one for themselves, though. It's exciting for them to watch
those seeds come up and to care for them as they grow.

➤ *Clean up the kitchen.* When you're washing produce, putting up preserves, and eating from scratch, you make a lot of dishes. You eat, you clean. It's that simple! Even the littlest ones can help with this.

➤ *Check on the babies.* There are always little animals around the farm: chicks, kits, kids, lambs, calves, or kitties. Put the kids in charge of checking on them. Are all of them there? Do they have food and water? They'll be naturally drawn to the babies anyway. Aren't we all?

➤ *Wash dirty farm gear.* Garden tools, buckets, wheelbarrows, or water troughs—all is fair game for a toddler with a bucket of soapy water and a scrubber!

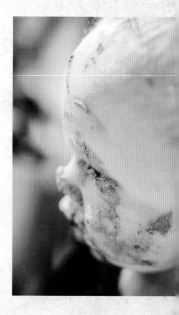

➤ *Harvest produce.* My kids are incredible tomato hunters. They're always finding the ones I miss! They're also great at digging potatoes. It's so much fun to share in the harvest with them!

➤ *Cook.* As soon as you can, get your little ones in the kitchen! Remember those eggs they carried in from the coop? Teach them to whisk some up into scrambled eggs. Teaching your kids to cook is one of the most valuable skills you can give them. And it all starts right here in the farm kitchen.

This past spring, my two oldest kiddos sat with me in the sheep pen while we watched a bottle lamb we had raised deliver her first lambs—a set of twins. "Little Lamb" and "Jacob" were born while we sat just a few feet away. Eleanor, the mother, lay there moaning, stretching, heaving (hey, I'm not judging. I've been there too). A variety of liquids squirted out and, naturally, there was blood. But my kids weren't scared. They were fascinated! Even when I had to pull one of the lambs because it was breech, they sat calmly in the corner and observed this entirely new world they were witnessing. "Mama! This is so special! I bet not everyone gets to see something so cool!"

True that, baby. And there's a lot more where that came from!

CHAPTER 3

For the Love of Chickens

Oh, chickens. What is it about poultry that makes homesteaders swoon? Say the word "chicken" to beginning homesteaders and instantly delightful images pop up in their minds. Does anyone else see beautifully colored birds gently scratching up bugs in a field of wildflowers? Or perhaps a basket full of perfectly clean blue eggs sitting in a vintage basket on the counter?

Well, it's sort of like that.

It's also sort of poopy, dusty, messy, and stinky. But hey, that's part of the fun, right? Regardless of the reality of raising chickens, backyard homesteaders will continue to be drawn toward this easy-care farm animal. When they give you breakfast each day for very little money, how could you not? Often regarded as the "gateway" farm animal, chickens pave the way for many homesteaders as they begin to venture into this radical way of life. Chickens can easily be raised on even the smallest amount of land, making them easy starter animals no matter what your experience.

We keep about forty laying hens in our chicken gang, along with some ducks, geese, and turkeys, all of which provide us with eggs. We also, somehow, ended up with a few

Chickens require very little from you and give the most beautiful gift each day!

roosters who manage to make their presence known, usually around 4:30 a.m. Our proudest, most boisterous rooster, Sir Henry, is a French Copper Maran with deep black and green tail feathers. When my little ones come with me to gather the eggs from the nesting boxes, I allow them to each carry a teeny little stick so that Sir Henry can remember who's in charge (most roosters learn pretty quick).

Then there's Susan, an older hen who was brought to our farm by another family needing to thin their flock. She's far too lazy to hop up into the nesting boxes and chooses instead to lay her eggs outside in an abandoned dog crate. "Susan's eggs" have become prized ingredients in our morning omelets (the kids are insistent that hers taste better). Big Bertha, our Giant Brahma, refuses to go inside the coop and will continually pace the edge of the exterior fence day after day, searching for an escape route. The Ameraucana chickens seem to have formed an alliance against the other hens and move as a gang through the chicken coop. And then there's poor Sir Isaac, my prized Lavender Orpington rooster whose mate was killed by the pigs (and who has since lost his place in the pecking order).

As you begin to build and care for your backyard flock, you'll no doubt notice the different dynamics of the birds and how they function as part of a teeny little society of their own on your farm. There's hardly a better way to spend an afternoon than sipping a cold beer, propped up against the chicken gate, watching your chickens live the good life.

Basic Requirements

Like all animals, chickens require a few basics for optimal health and production. There are as many variations as one could imagine, but the fundamentals remain the same: food, shelter, and water. Given these and a bit of care, your chickens will happily provide you with omelets for years to come.

Food Requirements

Because we eat our chickens' eggs, it's important to me that their diet is as natural and wholesome as ours. This includes a combination of whatever wild stuff they can forage around the farm, scraps from the garden, local

A whole-grain-based feed is ideal for your flock!

A Sneak Peek at What's in Our Feed Bag

Organic peas
Organic barley
Organic wheat
Organic corn
Organic linseed meal
Ground limestone
Oyster shell
Organic sesame meal
Fish meal
Vitamin and mineral pre-mix
Organic vegetable oil

grains, and (yes!) even bugs we grow right here at home. It may take a bit of time and tweaking to find out exactly what your chickens love the best. That's okay! Pay attention to their temperament, their likes and dislikes, and their laying patterns to track what works well for you and yours. Here's the most common feedstuff you'll find for backyard chickens:

Grain

Most chickens raised in backyards adhere to a largely grain-based diet. The crumbles and pellets available from your feed store are primarily grain-based, with the addition of minerals, vitamins, and usually an oil of some sort to bind it all together. In addition to crumbles and pellets, whole-grain feed is also available for chickens and is my grain of choice. New, local, family-run feed mills are popping up all over the country and are making it easier to source organic, local, non-GMO grains. Most grain is

Most pellets and crumbles are treated with high heat, which can destroy much of the nutritional value of the feed! When at all possible, source whole, cracked, or rolled grains.

Fermented Grain

Chickens love it when you ferment their grain! Their shells grow thicker and the yolks grow bigger! Fermenting their feed can also really help to save on feed costs, as chickens fed fermented grain tend to eat up to 75 percent less! To ferment their feed, simply cover the grain with clean water by an inch or two. Allow it to sit at room temperature for three days. Continue to start a new batch each day so that you have a fresh batch to feed the ladies each morning! A half-gallon or gallon-sized mason jar works well as a fermentation container.

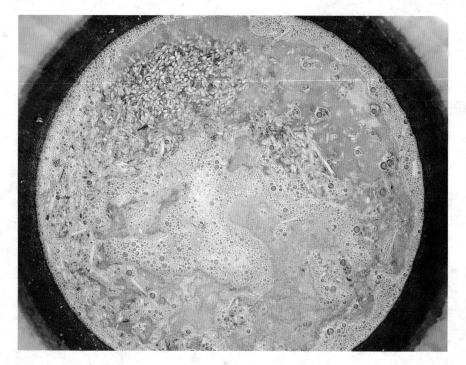

Soaking your whole-grain feed in water for a few days helps to make nutrients more readily available to your flock.

formulated to be between 16 and 18 percent protein and will often supply "extras" that the chickens like, such as oyster shells to aid in digestion and vitamins and minerals to maintain health.

Free-Range Feed

Wouldn't it just be great if all the chickens could just range free? They could eat bugs and grass all day long, roaming as they please. For some people, this is possible. They simply let their chickens graze around their yard as they wish. But for the backyard homesteader, this can present a few problems. For starters, chickens eat whatever they want, including your radishes and tomatoes! They also poop wherever they want. Like right on your back porch. Lovely.

The biggest perk of free ranging your chickens is that they'll source most of their feed themselves! Chickens are great at scratching up bugs and nibbling on grasses. This can drastically cut your feed bill. They're pros at controlling insect populations in the yard, as well as the garden.

There are always a few chicken gangsters that are bound and determined to escape and scrounge up some free-range food.

A few things to keep in mind when free-ranging chickens:

▷ Fence off anything you don't want them to eat.

▷ Fence off any area where you don't want them to poop.

▷ Fence off any area you don't want them to scratch up.

▷ Be prepared to hunt for those eggs. You'll be surprised at all the places they end up! Put your kids on the task . . . it's like Easter every day!

Lightweight bird netting can be highly effective in keeping your chickens contained.

A great way to free-range chickens without all that work is simply to utilize a chicken tractor (essentially a mobile chicken coop), which makes it possible to contain the birds, but also allows them to graze and scratch designated areas. Electric poultry netting can be another great way to contain the chickens, while still allowing them the freedom to roam. It will administer a slight shock to remind them to stay in the perimeter of the fencing. Simple bird netting can also be effective if you're looking to contain your birds from flying off.

Kitchen and Garden Waste

Chickens are wonderful at turning waste into eggs! In fact, our garden waste is so coveted it's always hard to decide who's going to get it–the chickens, pigs, and rabbits are all eager to snag a bite! I keep a large bucket on the counter at all times to collect kitchen scraps for them all. Chickens don't care too much for onions, garlic, or citrus, but otherwise are eager to clean up most anything we send their way. Don't worry about feeding something that will harm them. They will turn their beaks up at it if uninterested. Sharing our scraps and waste from both the kitchen and the garden not only gives the chickens a wider range in their diet, but also cuts down on our feed bill. A win-win situation.

Homegrown Protein

Grow protein on your farm for your chickens. Worms, flies, mealworms, and even meat scraps are all great protein sources for the birds and are easily grown at home. Kits for growing these bugs can easily be found online to get you started! Chickens are omnivores and will eagerly gobble up a variety of bugs and meats. After butchering, we often share some of the offal with our laying flock, and they love it! It's the ultimate form of recycling.

Mealworms can be easily grown at home and provide your chickens with extra protein!

Shelter Requirements

Chickens can be very forgiving when it comes to shelter, happily taking up whatever spare corner or space you've got in your backyard. All they require from you is protection from the elements and predators. Each chicken requires only a few square feet of space to roam, though the more space you can provide each chicken the better! Kept in a shelter that's too small, the chickens will begin to peck each other's feathers, develop bad attitudes, and are more susceptible to diseases. Also, kept too confined, they'll really start to smell. Seriously. Chickens are active little creatures and enjoy spreading their wings and wandering around, so allow them enough space to do just this. Happier hens mean healthier eggs.

To calculate the amount of space you'll need, count on at least six square feet of coop space per average-sized bird. So, if you'd like a flock of ten chickens, you'd need about sixty square feet of room in your coop (at a minimum). The more space the better. Ideally, you'll not only be able to provide your birds with space *inside* a coop, but also space *outside* a coop in the form of a run. This is a protected, outdoor location where the birds can soak up sunshine, scratch in the dirt and grass for bugs, breathe in lots of fresh air, and stretch their legs.

Just because you give your chickens nesting boxes doesn't necessary mean that's where they'll lay! Just go with it.

Be resourceful! There's no right or wrong way to build a chicken coop or house chickens. You're free to get creative. Some homesteaders prefer to start with prefabricated coops that are available online or through your local feed store. These are a great and easy option (though more expensive) for getting started and usually house three to seven chickens.

Homemade coops can save quite a bit of money but typically require a bit more sweat equity on your part. When we moved to our farm, the land had no chicken coop. It was the dead of winter, the ground was covered with snow, and we simply had to make do with what we had on hand. We sectioned off part of the shop that was on the property with metal paneling and hung bird netting along the opening to keep the birds from flying out. We (and by "we" I mean Stuart) then built a run using recycled cedar posts and 2-by-2-inch wire to create an outside run. All that was left to do was open the window that ran between the shop and the run and install a chicken ladder so that they could easily roam from coop to run. It's certainly not the picturesque coop you'd see in magazines, but it serves our flock well (and it's only a few more years until the honeysuckle and chestnut trees begin to fill in and create a special place for the ladies). They stay protected from predators, safe from the elements, have plenty of elbow room, and keep our family in an abundance of eggs!

Laying Boxes

Naturally, your chickens will require a place to lay their eggs. Prefabricated coops will come with laying boxes built right in, but if you're building your own, you'll need to include a safe place for them to lay. This can be as basic as a gigantic pile of straw in the corner of a run or as elaborate as a row of built-in, size-specific boxes. Generally, one box per three to four birds is sufficient. You'll often find that the hens will prefer one box on any given day and there will be a line of impatient hens waiting to lay their eggs there. Come on now, ladies. Wait your turn.

Inexpensive apple crates make great nesting boxes.

Predator Protection

Predator protection tends to take a super high priority with regard to your shelter choices, especially if you've ever come home to find a pile of feathers on your lawn. Raccoons, owls, coyotes, foxes, weasels, hawks, and even neighborhood dogs have an incredible ability to not only hunt out chickens, but also make serious efforts to break through your shelter's protection. A few small tips can help to keep your flock safe:

▷ *Get a guard animal.* Livestock guardian dogs, donkeys, and even geese are great protectors of chickens. Once trained, they'll serve you well and alert you to any intruders.

▷ *Lock the coop.* At night, chickens will naturally roost in their coop (or perhaps a tree, if they're free range!). If nighttime predators are a problem, locking the coop can be a great way to deter nocturnal hunters.

▷ *Keep a rooster.* If your birds are particularly susceptible to aerial predators, roosters can be a great way to alert the hens when danger is nearby so they can find shelter. The rooster has a natural instinct to protect his ladies. What a gentleman.

▷ *Reinforce everything.* Think your coop is strong? Make it stronger! Predators want to get into your coop even more than you want them to stay out. Thicker-gauge wire, longer screws, rock barriers, or stronger lumber–bring it full force!

Common Breeds

There are many common types of chickens that you'll find at your local feed store each spring, but if you hunt around for a specialty breeder or hatchery, you can often find some unique breeds as well. Your goals as a chicken keeper will likely sway your decision as to what breed you'll bring home. If your goal is maximum egg production, white leghorns are a great option. However, if part of your goal is to actually *enjoy* your chickens, then leghorns may not be your best bet because frankly, leghorns are crazy ol' things. Regardless, it's pretty hard to go wrong. There are hundreds of breeds available!

The Egg

Okay, not to be crass, but one of my dear farming friends calls these incredibly delectable orb-shaped treasures "butt nuggets" and, well, it's kinda true. Yes, that's sorta where an egg comes from (always important to know these things), but they're delectable, nonetheless. Eggs are the reason that most of us begin backyard farming. We want those home-grown, deep yellow yolks on our breakfast table. We want those little trophies of chicken farming lined up in a carton on our countertop. We want eggs from chickens that we've watched work through our kitchen scraps, scratch around for bugs and grassy bits, and greet us at our backdoor in the morning. Eggs are an edible testament to what we love about backyard farming.

Eggs only come from female chickens, also called hens. Male chickens, called roosters, cannot lay eggs (and yes, I've been asked that question more times than I can count). A hen does not need a rooster to produce

For eggs, consider these breeds:
Rhode Island Red
Orpington
Leghorn
Plymouth Rock
Sussex
Maran
Ameraucana
Black Star
Barred Rock
Wyandottes
For meat, consider these breeds:
Freedom Rangers
Cornish
Orpington
Australorp
Red Ranger
Pioneer

Eggs are even prettier when they're collected with garden flowers on the way back from the coop.

"The black chicken," as we like to call her, is a French Copper Maran and a beautiful layinghen.

eggs. The eggs will come no matter what! However, if at any point you'd like those eggs to be fertilized so that a hen can hatch chicks, then a rooster is necessary. This is basic knowledge one must have if one is to keep a laying flock. Aren't you glad I'm here to fill you in on such complicated and important information?

A Hen's Life Cycle

On average, a pullet (a hen that is less than a year old) will begin laying eggs at around twenty-two weeks of age, though the eggs will be slightly smaller than normal. I love getting those first teeny little eggs! It will take her about eight more weeks to begin laying normal-sized eggs in full force. All pullets and hens lay eggs naturally, but the amount of eggs they will lay is determined by breed, genetics, and various environmental factors. A

typical laying hen in her prime will lay about one egg per day. And while the hen's "prime" will only last about two years before her egg production drops, she can still continue to give you many eggs throughout the rest of her life. If prime egg production is a goal with your backyard flock, you may consider culling birds that are more than three years old and giving them a second chance to contribute to the stew pot. If you're happy to keep some freeloaders along for the ride, they can continue to bless you with sporadic laying for a few more years. I tend to keep my ladies around well beyond their prime. After years of faithfully feeding our family, I figure they've earned their retirement. Each year, we raise up new chicks for the backyard flock so that we have a steady supply of new layers in high production.

Expected Production and Environmental Factors

As I briefly mentioned above, what you can expect your laying hens to produce is determined by multiple factors. Some of these you can control, such as light exposure and temperature, and some you cannot, such as genetic makeup. Each hen is unique. On average, a strong laying hen will give you about 250 eggs in her first year of life alone. While you can't change the genetic makeup of a hen, tweaking a few environmental factors such as the following can increase and optimize your egg production:

Light

Just as all things in the natural world, egg production is affected by day length (read, light!). Light triggers a hen's pituitary gland to produce eggs and thus, a lack of light produces the opposite effect. Egg production tends to taper down, or shut off completely, during the short, dark winter days. On average, a hen requires around fifteen hours of daylight for optimal egg production. Supplemental fluorescent lighting to reach the desired day length can easily be added to the coop during the fall and winter to maintain production, with no side effects to the hen's health. More light equals more eggs. And more eggs means more omelets. Am I right?

Temperature

Both extreme heat and extreme cold will cause your hen's egg production to decrease. Can you blame her? I don't want to be productive in either of those temperatures either! Moderating the temperature in your coop can help to boost production. In the heat of the summer, this can be in the form of shade (always necessary!), ice blocks, and fans. In the winter, a heat lamp, draft-free coop, and thick bedding can help to take the edge off the cold. Hens lay best when the temperature is between 50 and 80 degrees F, and while it's nearly impossible to maintain this ideal range year-round, a little bit of help from you can surely affect the amount of eggs that make it to your basket.

Stress

Happy hens lay eggs. Stressed-out hens lay eggs too, but not nearly as many! Maintaining a safe, calm, and stress-free environment will increase the overall production of your flock. If the neighborhood kids chase your hens around with sticks, you can expect egg production to drop for a few days. (Why is it that kids who don't have farm animals always want to chase them around with sticks?) The moral of the story is this: Keep the hens calm and happy!

Egg Collection

Ideally, you'll gather the eggs from your nesting boxes each day. Eggs can easily become cracked and dirty when hens are continually walking on them, lying on them, or (gasp!) even pecking at them. Truthfully, gathering eggs has to be one of my greatest joys on the farm. It's how we taught our children to count ("one egg, two eggs, three eggs . . .") as we gently tucked them into our vintage basket. If I ever grow tired of seeing that beautiful pile of eggs in the nesting box, I'll quit being a farm girl. But I seriously doubt that'll ever happen.

Also, let's be honest here. Sometimes I forget to clean out the nesting boxes and instead of seeing a beautiful pile of multicolored eggs, I see a bunch of poop-smeared eggs. And sometimes, instead of gathering

Sometimes eggs come out bloody. Ouch. Sorry, hen! Don't worry . . . after a quick washing, they're still fine to eat.

them in my vintage basket, I shove them in my winter coat pocket or in a random dirty can that's been lying around (and most likely held compost scraps or dog food).

Just keepin' it real.

Egg Storage

To refrigerate or not refrigerate–that is the question! Fresh eggs that have not been washed do not require refrigeration. Did you know that eggs are laid with a natural, antibacterial coating? This is called "the bloom" and completely closes all the pores of the egg. It's like a force-field of natural awesomeness. When eggs are washed, this coating is removed. Pores are opened. And bacteria have the perfect opportunity to strike. Bam!

HOW TO TELL IF AN EGG IS FRESH

If I'm questioning whether or not an egg is good, there is an easy and quick way to check! Fill a small glass with water and put the egg into the glass. A fresh egg will sink to the bottom. A semi-fresh egg will sit on the bottom but point upward. A bad egg will float. Here's an even better tip: Don't eat the floating eggs!

Contamination. Therefore, if you do decide to wash your eggs, you can't keep them at room temperature any longer. Bacteria rules dictate. (USDA rules also dictate that eggs cannot be sold outside of refrigeration so even the swankiest health-food stores will always stock their eggs in cold storage.)

In fact, until just a few decades ago, room temperature is how people kept their eggs and how many people still keep their eggs around the world. Americans generally insist that our eggs be refrigerated (probably because the quality is so poor we have compensated somehow . . . but that's another topic!). Store-bought eggs that are washed, bleached, and then have a synthetic coating applied can be stored at room temperature, but only for a short time (in comparison to farm-fresh eggs that have not been washed and may store well at room temperature a week or more!). Commercial eggs are laid under terrible conditions and the threat of contamination is much higher than with a few chickens on a farm.

Unwashed, farm-fresh eggs will last about thirty days in the refrigerator or about a week on the counter at room temperature.

Recycled cardboard egg cartons or baskets will hold your eggs just fine until you're ready for them, whether you decide to keep them on the counter or in the refrigerator.

Egg Preservation

Though not a super-common practice in our day and age, egg preservation is still a valuable skill to have if your ladies go on a laying strike or if you've got an abundance of eggs to deal with. Stick these methods in your back pocket for future reference:

Freezing. Eggs freeze fairly well, whether in ice cube trays to keep them separated or already whipped into a scramble. This is a great way to keep eggs for baking or using in recipes–just be sure you write how many are in each container when you freeze them so there's no guessing when you take 'em out of the freezer (points to self).

Pickling. Oh, yes. Just like you've seen in gas stations, but so much better when they're made from your own eggs!

Pickled Eggs

12 hard-boiled eggs, peeled

2 cups vinegar

1 teaspoon salt

1 large onion

⅓ cup dehydrated whole sugar cane

1 tablespoon pickling spices

1. Place the hard-boiled eggs into a canning jar.
2. Combine the remaining ingredients and boil them in a small saucepan for 5 minutes.
3. Pour this mixture over the top of the eggs and place the lid on the jar.
4. Leave the jar on the counter overnight before placing it in the refrigerator. The eggs will keep in the refrigerator for a few months.

Coating. A very traditional way to preserve eggs, coating involves coating an egg in mineral oil within twenty-four hours of it being laid by painting the oil onto the eggshell. This oil coating helps to prevent evaporation when you store the eggs in a cool, dark place. The eggs should be stored around 55 degrees F and no higher than 75 percent humidity. They'll keep for up to three months. Egg storage victory!

Egg Washing

Poopy eggs happen—especially in the winter when everything is wet, sloppy, and frankly, often gross. During those times of year, I do wash the eggs. Because I prefer my omelets sans poop, please.

Dry Method

1. Get yourself a sponge, super-fine-grain sandpaper, or other scrubber.
2. Scrub the poop from the eggs. This method leaves the bloom intact, which means that the egg can still be kept at room temperature. However, some eggs are too dirty for the dry method, which works best for smaller, drier bits of poop.

Wet Method

1. Fill a bowl with *warm water*. Cold water will cause the bacteria on the surface to be driven into the egg. Please, don't use cold water. Just don't.

2. Place the dirty eggs into the warm water, only for a few seconds, then utilize your sponge or a wash rag to gently scrub the poop from the shell. This should be pretty easy, unless a hen has broken an egg over the other eggs. Dried egg is like cement. But poop is pretty easy to remove. See what sort of valuable information I bring to your life?

3. After washing the egg, place it on a towel, rub it dry, and place it in a bowl or a carton. Washed eggs have lost the bloom and must be stored in the refrigerator. I always use them before cracking unwashed eggs because the shelf life decreases significantly once they're washed.

Homemade Egg Pasta

There's hardly a greater joy in the kitchen than fresh pasta made with your very own farm-fresh eggs. It's a wonderful way to use up extra eggs and comes together in just a few minutes. Don't be intimated! You can do it.

2 cups unbleached, organic, all-purpose flour (for heartier dishes, sprouted flour or einkorn flour is a great alternative!)

4 eggs

1. Pile the flour onto a clean work surface. Use your fingers to make a well in the flour.

2. Crack the eggs into the flour.

3. Use your fingers to slowly incorporate the eggs and flour together. It'll be tough and crumbly—that's okay. Just keep workin' it. If needed, add a teaspoon of water at a time until the dough forms a tight, barely sticky ball.

4. At this point, you can divide the dough into two balls and roll it out into sheets with a pasta machine. Alternatively, you can roll it out using a rolling pin on a floured work surface. Be prepared to put some elbow grease into it if you're hand-rolling! It's springy and opinionated, that dough. Dust the pasta sheets with a bit more flour if you're having problems with the dough sticking to your machine or countertop.

5. After rolling the dough out to the desired thickness, you can bag it up in a plastic bag or wax paper and stick it in the refrigerator to rest until it's supper time. You can also utilize your pasta machine to cut it up into whatever noodle shape your little heart desires—most come with the spaghetti and linguine attachments.

Chicks can be born or brought to the farm each year to replenish the aging laying flock and to provide you with more eggs!

Raising Chicks

Each spring, you'll no doubt find a metal tub of little peepers stashed somewhere in my home to replenish our laying flock with young birds. Because we've never had a hen that has successfully hatched out her own eggs (more on this below), we've been dependent on a local chicken farmer or feed store to supply us with chicks or fertilized eggs each year. Chicks require little more than food, water, shelter, and a warm environment to thrive, but, let's be honest, raising chicks can be tricky even under the best conditions. All baby animals are fragile, and chicks are no different, often finding creative ways to die. But don't be discouraged. It happens to nearly everyone. Just plan on ordering a few more than you need and account for a small amount of "shrinkage" in your new flock.

Hatching in an Incubator

Some of you may be so bold as to hatch your own fertilized eggs in an incubator! This is a fantastically fun experience that I hope many of you get to try at least once in your backyard farming life. After twenty-one days in an incubator baby chicks start "peeping" out through their shells. This is a great way to bring in new breeds to your flock that are a bit more specialized and unique. This method requires commitment on your part and some specialized equipment (I trade vegetables for the use of a friend's incubator). In order to hatch, the eggs must be kept at a steady 99.5 degrees F for twenty-one days. For the first eighteen days, the humidity must also be kept around 45 percent. In the final three days before hatching, the humidity must be increased to around 70 percent. Most fertilized eggs from a farmer will come with detailed instructions on how to get the best results from incubation.

Hatching fertilized eggs in an incubator is a fun way to welcome new chicks into your flock!

"Big Bertha" as a baby was still big for a chick!

Buying Chicks

In the spring, feed stores are filled to the brim with little chicks! I know because it's a time when I seriously have to practice willpower (ah, who are we kidding?). These chicks are typically only a day or two old when you get them and take away the stress of having to hatch your own. But because they come from large hatcheries, they're usually limited to a few specific breeds.

Leaving it Up to the Mama

Some hens experience something called "going broody," in which they go into a trance-like state while sitting on a pile of eggs for the twenty-one-day incubation period. During this time, they protect the eggs and keep them at the necessary temperature and humidity levels. After the chicks hatch, the hen will take to mothering the little ones and show them the ways of the flock. This is the most natural and easiest way to add new chicks to your poultry crew. Unfortunately, after decades of selective breeding, a hen's

Our turkeys tend to be more broody than our chickens. Ever gone near a broody turkey? Yeah, me neither.

maternal instinct isn't a guarantee, so it can be hard to find a hen that will make a good mother (or will go broody at all!).

If you don't want more chicks or don't have a rooster to fertilize the eggs, having a broody hen is rather inconvenient! Not only will she stop laying eggs during her broody period, but she can also cause other chickens to go broody. That's a whole lot of hormones in the coop.

Food Requirements for Chicks

Chicks do best with a slightly higher-protein feed until they lay their first egg. Their bodies simply require more protein and less calcium to grow best. Chick starter feed is available at any feed store. Naturally, they should have free range of feed at all times.

Shelter Requirements for Chicks

I've kept chicks in everything from cardboard boxes to metal tanks to bathtubs. Add in a few inches of bedding and you're good to go (just make sure they're protected from dogs and three-year-old farm boys). It doesn't take long for the chicks to learn how to fly out of their containers, so a bit of chicken wire or netting over the top will ensure they stay where they're supposed to. It's ideal to supplement the temperature with a heat lamp, as chicks do best at 95 degrees F during their first week

Breaking a Broody Hen

Remove her from the nesting box multiple times per day.

Put her in a wire-bottomed cage, with only food and water, until she lays an egg.

Block off the nesting boxes.

Bribe her out of the box with her favorite treats. Hey, chickens like treats too!

Broody Hen Facts

A hen going broody is determined by instinct and hormones.

The hen will quit laying eggs while she sits on the eggs.

The eggs must be fertilized by a rooster in order to hatch.

Broody hens tend to be territorial and protective of their eggs.

Proceed with caution!

A NOTE ON PASTY BUTT

Chicks are susceptible to "pasty butt," which is when their poop clings to the feathers on their backside and builds up, preventing them from . . . well . . . pooping. If you notice a chick with pasty butt, gently wet its backside in warm water until the clump can be removed, dry it off, and then slather a bit of coconut oil around its vent (the hole under its tail where the excrement and eggs come from) to prevent this from happening again. A tablespoon of raw apple cider vinegar in the chicks' waterer can be extremely helpful in preventing pasty butt.

of life, after which you can reduce the temperature by 5 degrees each week (week one: 95 degrees F, week two: 90 degrees F, week three: 85 degrees F, and so on). A regular ol' heat bulb and reflector from the hardware store will do the trick. It's also helpful to learn how to "read" the chickens. If they're huddling together under the light, they're cold. So, increase the temperature. If they're hiding in the corners as far away from the light as possible, they're too hot. Pull the heat lamp back. Read the chick. Know the chick. Be the chick.

Transitioning Them into Your Flock

Once the chicks have grown in their "real" feathers and are accustomed to outside temperatures, they can make the transition to the big-girl coop. Naturally, as you release the chicks into the coop, the older hens will want to establish the pecking order. This may mean your little ones take a bit of a hit their first few days. Typically, they'll just run away and quickly submit to the already-present pecking order of the coop. If they decide to challenge the system, it won't be until they're older. But still, it's best to keep an eye on them their first few days. Chickens can be cruel. I like to keep my chicks in an enclosed cage in the coop for a few days so they can get used to their surroundings and the other hens before sending them out into the wild world.

Why Stop at Chickens?

There's a beautiful world of birds out there just waiting for you to explore! While we started with chickens on our farm (as many farmers do), we quickly expanded to other feathered creatures such as ducks, turkeys, and geese. They're simply too much fun to be without.

Ducks

Yes, ducks are awesome. I love watching them waddle around the coop. We keep ours right alongside our chickens and give them the same feed, the same nesting boxes (though they tend to lay *under* them), and a little pool for swimming around in. Just so you can't say I didn't warn you, duck poop is smellier than chicken poop. Ducks will provide your farm with wonderful, large eggs that have up to double the nutritional value of a chicken egg! The duck egg also has a thicker shell, causing it to stay fresher longer.

Turkeys

Our first turkey was named Thanksgiving. Each morning, when I milked Sal the cow, he would ride down to the barn with me on the golf cart before following me into the milking parlor, where he would eat loose oats off the ground that spilled out of the cow's feed bucket while I milked. He would linger for a while longer and keep us company before following me back up to the house to join the flock for the remainder of the day. Since Thanksgiving left us for . . . well, the supper table, we've added three more turkeys to our flock. They're large and slightly awkward looking, but a joy

to have around nonetheless. The hens leave me delicious eggs each day and faithfully go broody in the spring. The tom turkey is only here to breed the ladies, plus he's funny to look at, so he can stay. We all need a bit of extra laughs in our lives, right? Get yourself a turkey.

Geese

It took a bit of convincing to make my husband realize how badly we "needed" geese on the farm, but as all the animals eventually become, our Pomeranian geese are now part of our family. They happily waddle around the pasture, nibbling grass and swimming in a little water trough. The best part about geese (besides the ridiculously delicious eggs) is their natural protective tendencies. Our geese are as good as guard dogs—eagerly letting us know if a car is coming down the drive or the neighbor is coming over for a visit.

Also, geese are gigantic. You haven't lived until you've had a goose chase you around the pasture with its wings spread as it honks and hollers at you. See how much excitement you've been missing out on all these years?

Commercially raised geese are fed (often force fed) a high-fat and grain-based diet. What does nature have to say about that? The opposite. Geese eat primarily grass, sprouts, roots, and leaves, supplementing lightly with insects.

Poultry Pests

Chickens, and fowl in general, are susceptible to a variety of pests that'll surely make your skin crawl. The first time I found one of my hens with chicken lice, I'm pretty sure I almost scratched my skin off out of sheer disgust! But since that time, I've come to accept that some level of infestation is normal. Here's what you should watch out for:

Lice

Chicken lice bury themselves in the chicken's feathers and will drive her absolutely bonkers. They have six legs and look somewhat like fruit flies. They are light brown in color and tend to congregate around the chicken's vent and eyes. They feed on dead skin cells and will lay their eggs at the base of the feather shaft, which will look like heavy, white buildup where the feather meets the skin. Left untreated, the lice can actually kill the chicken, so take an infestation seriously and treat it right away!

Northern Fowl Mites or Red Roost Mites

A mite infestation is just about as bad as it gets in the poultry world. I know from experience. These mites range from light brown to red and tend to hang out in the chickens' nesting boxes and roosts. They hide in crevices of wood during the day and come out at night to bite the chickens. Mites also tend to congregate around the chicken's vent–an easy place to spot them if you're checking your chickens. The mites can eventually kill the chicken (she will die from anemia from all their little bites!) so treatment is key.

Scaly Mites

Scaly mites are these really freaky little monsters that actually burrow up under the scales on the chicken's legs and dig little tunnels, naturally causing lots of discomfort for the chicken. This will eventually cause little scales to build up on the chicken's legs that look white, flaky, and disproportionate. Unlike the lice and mites listed above, the scaly mite doesn't threaten the life of the chicken but can cause the loss of toes, disfigurement, and severe discomfort.

Clean bedding keeps the hens happy and pests at bay.

Signs of Lice and Mites

Pale combs and wattles

Itching

Bald spots

Decreased egg production

Lethargy

Redness or scabs around the vent

Creatures crawling around on your chicken (obviously)

There are a few preventative measures that you can take within your coop to keep pests to a minimum:

- *Keep it clean.* One of the best ways to avoid these poultry plagues is to keep the coop clean! Ideally, the coop should be cleaned of old bedding and waste on a weekly basis. Old, gross bedding is the breeding grounds for these nasty beasts. If an infestation does occur, remove everything you possibly can from the coop. Strip it down and pressure wash it all, paying close attention to cracks and crevices in the wood, laying boxes, roosts, etc. Clean it as thoroughly as you possibly can.

- *Early detection.* Chances are, if you have poultry, you'll encounter pests at some point. But bimonthly "chicken checks" will help you to catch any sort of infestation early on, and this is a great form of prevention for the rest of the flock.

- *Dust baths.* Give your chickens access to a dust bath! We use a combination of sand, diatomaceous earth, and wood ash in a small plastic tub. The chickens hop in, roll around, and dust themselves, working the dust deep down into their feathers. This will help to kill and prevent lice and mites. There are many mixed opinions on using diatomaceous earth around the chickens, as it can cause some respiratory distress. I'd still rather deal with possible respiratory distress than pesticides, though!

- *Garlic oil.* A healthy glug of garlic oil, or crushed garlic cloves, mixed into the chickens' feed will make their blood taste of garlic! Mites don't like this. Without a host to eat from, the mites will die. Die, mites, die! Garlic oil can also be sprayed directly on the mites.

- *Garlic water.* Yes, you can even crush some garlic cloves and put them in the chickens' water! Anything to make the chicken smell and taste of garlic. Die, mites, die!

- *Keep 'em healthy.* Fermented feed, access to grass and fresh air, and raw apple cider vinegar in their water are all great ways to keep chickens healthy. The healthier the birds are, the less susceptible they are to pests.

▷ *Neem oil.* Neem oil can be a wonderful way to treat mites that are in the wood of the coop. It will coat and suffocate them. Spray it everywhere. And then the next day, do it again. And again. And again. Until you don't see any more!

▷ *Petroleum jelly.* For scaly leg mites, petroleum jelly slathered very thickly over the legs of the bird can help to suffocate and kill the mites. This should be repeated every day until the legs are healed.

As gangster as they can be at times, I can't imagine our farm without chickens. It's become such second nature for me to look out the window to see what's going on in chicken land. Their squawks are always welcome (yes, even the rooster's morning wake-up call!) and there's hardly a better sight than seeing them scratch around the pasture. Chickens bring life to the farmyard. They bring joy to morning feedings. And they bring eggs. And Amen.

CHAPTER 4

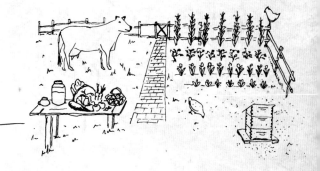

The Family Dairy

As you know now, the very first animal I bought for our farm was a dairy cow. I had no idea what I was doing, I didn't have a pen set up, and I didn't even know how to milk a cow. Heck, I wasn't even living in the same state as the cow when I purchased her! I've learned a lot since those early days–lessons learned through the blood, sweat, and tears that can only come from steep learning curves.

But despite the trials, I also experienced a level of love that I didn't know could exist on a farm. I've never bonded with an animal the way that I have with our dairy cows. Naturally, as you're hunkered in under the animal's flank, all up in their business, you bond. You spend your mornings and evenings together, rain or shine. You spend time together even if you don't feel like it. Even if one of you is in a grouchy mood (yes, animals get grouchy too). Come hell or high water, it's you and your dairy animal. True luv 4ever.

It goes without saying that a dairy animal is a commitment. Lactation is constant. When an animal is lactating, she's always got that milk flowing. That means you need to milk her every day, twice a day. If you want to skip out of town for a rodeo on the weekend, you better have a farm sitter come and take care of the ol' girl. Come milking time, she'll be bellowing from the pasture letting you know she's ready.

All of us mamas who have nursed our own babies can understand. There's no pain like engorgement pain! Sorry I said *engorgement*. But it was the only way for me to get the point across to you: A dairy animal needs to be milked on schedule, all the time.

While they require a bit more commitment on your part, dairy animals also give more than almost any other animal on the farm. Our dairy cows provide us not only with fresh raw milk and cream but also the ability to make butter, yogurt, kefir, sour cream, and a variety of cheeses! Once you get comfortable with the routine of milking, it can easily be done in less than twenty minutes. That's a pretty small commitment for such a huge (and creamy!) payoff. (Bonus: Dairy animals come in a variety of shapes and sizes to fit your farm! Whatever size that may be.)

The Basics

Our sweet heifer Cecelia will be a new milker next year.

Before we dive into specifics, let's cover a few basics. Yes, your dairy animal needs to give birth in order to produce milk, just like a human. And no, you can't milk a male. I've been asked that question more times than I care to count.

After a dairy animal gives birth, she begins to produce milk for her babies. This is called "freshening." During the first few days of freshening, the animal will produce colostrum. This is a thick, sweet, nutrient-dense milk that is incredibly beneficial in getting the baby off to a healthy start. It kick-starts the baby's digestive system and prepares its gut bacteria for digesting milk. Some dairy animal owners will allow the babies to drink this colostrum before they start milking while others will milk the colostrum and keep it in the freezer, in case they need it in the future for emergencies (such as if an animal dies giving birth). That bit is up to you. Just know you won't get actual milk until day three or four. You'll know by the density and the taste: Colostrum tastes like sweet and salty minerals. Ahem. Or so I'm told.

But what about you? What sort of dairy animal is right for your farm? Let's take a peek at some determining factors and pros/cons of common dairy animals.

Dairy Cows

Obviously the largest of dairy animals, a dairy cow will give you the most milk. We've always kept Jerseys, and on average we milk out more than four gallons a day from one cow! That's a lot of milk, baby. Dairy cows are the "queens of the farm" and tend to rule the roost, because they're *that* wonderful. It's important to consider how you will breed your cow before committing to her. Will you take her to a friend's bull? Will the bull come visit your farm (note: check your fencing)? Will your veterinarian or an A.I. tech artificially inseminate her? Cows are lovely but can also be expensive lawn ornaments if they can't get bred. A cow's gestation is nine months, meaning you'll need to think ahead about when you should breed your dairy cow and in what month you want her to freshen. We usually shoot for May calves, which means breeding in late August or early September. On average, a cow will give you one calf per year (twins are rare). She will need to be rebred three months after giving birth. Don't worry, you can milk her while she's pregnant! Just give her a few months dry (meaning she's no longer producing milk) right before she calves so she can put all her energy into the new calf.

Pros: Cows produce more milk than any other dairy animal. And more cream. And Amen.

Cons: Cows eat more than other dairy animals and require more room to roam. They are larger in size and thus, not as many people are comfortable working with them.

Hiro the bull comes to make his rounds each summer.

Sample Annual Cow-Milking Calendar

Breed in August

Calve in May (nine months later)

Rebreed again in August (three months into milking)

Stop milking sometime in February

Cow is dry until calving again in May

Common Dairy Cow Breeds

- *Jersey.* My drug of choice, I'm keen on the Jersey's high milk fat and those beautiful brown eyes. Our Jerseys have been incredibly sweet and easy to manage. Jerseys are a British breed, hailing from island of Jersey in the English Channel, and are known for their grazing ability, heat tolerance, smaller stature, feminine characteristics, and high production. A full-grown cow weighs in at about 900 pounds.

- *Brown Swiss.* Imported from Switzerland, these beauties are often known as the gentle giants of the cow world. Weighing around 1,300 pounds, they're obviously much larger than a Jersey. They are popular for their quiet dispositions and easy handling.

- *Holstein.* The most recognizable of all the dairy breeds, Holsteins are the traditional black-and-white cows that we've all grown to love. They're gigantic. And boy, do they give some serious milk! Because of their incredibly large size and production (on average, about nine gallons per day!), they're not typically the breed you'll find on a small farm but rather at commercial dairy operations.

➤ *Dexter.* I've got a soft spot for Dexter cattle, as we've bred our cows to a Dexter bull and always end up with the sweetest little Jersey/Dexter calves. Dexters are a dual-purpose breed, raised for both meat and milk, making them wonderful candidates for the backyard farm. Dexters are small, with a full-grown cow only averaging about 750 pounds, but they can still produce up to 2½ gallons per day. They're much less expensive to feed, due to their smaller size and lower production, making them a wonderful gateway bovine breed.

Dairy Goats

Much smaller than cows, goats can still deliver a pretty hefty supply of milk. If you're limited on space, or if the idea of tucking into a thousand-pound cow's flank to squeeze an udder the size of a beach ball is slightly intimidating, then goats may be just what you need. You can boss 'em around and really teach 'em who's in charge. At least, that's what I'm told. I, myself, am not a goat person. What can I say? I'm partial to bovines. My close friend DaNelle, on the other hand, is *the* dairy-goat lady. I've been listening to incredible milking stories of her goat herd for the last few years so, dare I say, I'm completely qualified to boast of their awesomeness. Goats average a five-month gestation period and are typically bred naturally with a buck.

Pros: Obviously smaller than a cow (an average doe typically weighs in at around 130 pounds), goats take up much less space and eat significantly less, making them much easier for the backyard homesteader to maintain.

Cons: Goats tend to live up to their reputation of being escape artists. Containment can be a real challenge, so prepare yo'self!

Common Dairy Goat Breeds

➤ *Alpine.* Alpine goats originated from the French Alps. Their agility and adventurous nature are proof of that! An average-sized breed, Alpines have horns, erect ears, and no distinct coloring. They thrive in all climates and produce up to a gallon of milk per day!

- *Nubian.* Nubians may just be the most common dairy goat around. They are gentle, have mild-flavored milk, are known for their spunky personalities, and have those deliciously cute floppy ears.

- *LaMancha.* The first time I saw a LaMancha goat, I thought it was a joke. They're earless! Ha! Easy to pick out among a herd, the LaManchas are known for their calm and sweet demeanor.

- *Nigerian Dwarf.* As its name suggests, the Nigerian is a smaller goat that is noted for its incredibly rich and sweet milk, making it a great option if space is limited! Nigerians will give about a third of the milk that an average-sized breed will.

Dairy Sheep

I'll be honest, I've always secretly (okay, not so secretly) wanted to own a sheep dairy. I can't help myself—I just love those little furry creatures! Though they may not be the first to come to mind when you think of dairy animals, they've still earned their place among the ranks by providing some of the most decadent milk available. Sheep milk is prized for its complex flavors and sweetness—hello, cheesemakers! (Bonus: they can grow fleece {read: wool} at the same time.) Sheep average a five-month gestation period and are most commonly bred naturally with a ram.

Pros: Multi-purpose sheep can provide your homestead with meat, milk, and wool all at the same time. A winning combo!

Cons: Also notorious escape artists, sheep can be slightly more finicky than other farm animals, as they tend to succumb to illness a bit easier than others.

Common Dairy Sheep Breeds

- *East Friesian.* The most common dairy sheep, East Friesians are the Holsteins of the sheep world. They're super heavy producers, have good personalities, and are well adapted to commercial dairy setups.

FAT TAILS VS. RAT TAILS

"Fat tail" refers to sheep that have been bred for large hindquarters and thus, large tails. They look slightly disproportionate with their gigantic hind ends and tails that can reach up to a foot wide! "Rat tail" refers to sheep that have an average, thin tail, which may or may not be docked, depending on the shepherd's preference.

▷ *Awassi.* Found primarily in the Middle East and Asia, the Awassi breed can be used for both dairy and meat purposes and is well adapted to harsh climates. They are a fat-tail breed.

▷ *Lacaune.* The most common dairy sheep utilized in France, they're the magic behind Roquefort cheese. Though they produce slightly less milk than, say, an East Friesian, they produce slightly higher milk solids (read: fat!), which is great for cheesemaking.

What You Should Expect

Before we get started on the how-to, let me set a few expectations for your first few milking sessions:

1. Expect to feed your first few milkings to the pigs or chickens. Your milking bucket will contain hair, hay, and poo particles from your lack of ability to quickly milk and the animal's lack of patience with you—a quick tail-flick or kick is all it takes to get nasty stuff in your milk!

2. Be ready to practice patience. When I was learning to hand-milk, my cow stuck her foot right in the bucket. As I was tromping up to the house to wash it out, Stuart hollered at me, "How's it goin' down there?" "Bad! Really, really bad!" I hollered back. And then I said a swear word that I'm not very proud of so I won't type it. But I was

frustrated! And angry! How dare she stick her foot in my bucket of milk? What is she, an animal? Oh wait . . .

3. Get comfortable with your animals. They need to know your smell, know your purpose, and know your presence. The more time you spend in their company loving on them and grooming them, the more comfortable they'll be with you milking them.

4. As you're learning, always take out an extra milk bucket. Just trust me on this one.

5. Pour your milk into another clean bucket frequently. Chances are, as you're learning, your animal will kick at you. Dumping the milk frequently into another container will prevent you from losing all your milk, should the animal (literally) kick the bucket.

6. Expect it to hurt. The first time I ever milked a cow, I was almost in tears. My hands have never known that kind of pain! But by the second milking, the extreme muscle cramps were gone, and my arms

already felt stronger. On the plus side, I no longer need to work out my arms and shoulders at the gym (oh wait . . . I never did that anyway . . . never mind). Embrace the burn, my friends.

7. Have a second milker on hand. Until you get proficient at it, having some emotional and physical support is huge. Have a backup.

8. Prepared to be blown away by the taste. Even though we've had an excellent source for our raw milk, after I tasted our milk the first time, I was shocked. It was so incredibly sweet! The sugar in the milk tends to break down quickly, so the fresher the milk, the sweeter the taste.

9. Don't give up. As with most homesteading skills, the first go-round can be a little hairy. But that's okay! The important thing is not to quit. If I gave up gardening whenever I lost a crop to bugs, where would we be? If every time one of our chickens was attacked by a predator we gave up raising them, where would we be? If every time our cow was saucy we gave up milking her, where would we be? Out of food. That's where we'd be.

Milking Setups

Because we got our first milk cow before we had a "proper" setup, we learned to milk by simply tying her up and sitting next to her. I love that this method eliminates the need for fancy equipment. This method tends to work better for cows because they're larger and you can easily tuck in under their flank. For goats, on the other hand, you'd have to crouch down low and this would, I imagine, lead to a slightly sore back and neck at the end of the day.

A basic platform called a milking stanchion will serve you well for milking those little critters that are too teeny to milk otherwise by raising them to a comfortable height. Stanchions are typically built with a head catch that secures the animal during the milking process. If you want to get fancier, you can add a little bucket at the bottom of the head catch from which the animal can feed. Lucky animal.

How to Milk

1. Grab the animal's teat where it meets the udder with your thumb and index finger. Gently squeeze to trap the milk in the teat.

2. While maintaining the pressure with your thumb and index finger, use your remaining fingers to tighten down the teat, first with your middle finger, then your ring finger, and finally your pinkie. This pushes the milk farther and farther down the teat until it's finally squeezed into the bucket! Release all your fingers so that the teat can fill back up with milk. Repeat. Don't worry. You'll get faster.

Home-Dairy Musts

Dairy animals require a bit more attention than some of your other farm animals, but ultimately, it all comes down to a few basics. Follow a few steps and you'll reap the rewards from your animal for years to come.

▷ *Maintain a healthy animal.* I know this probably sounds like common sense, but it's important. Sick animals will not produce healthy milk. Your animal should have access to fresh water, fresh grass, high-quality hay, clean living conditions, and mineral supplements or salt licks. Because you'll be seeing your animal multiple times a day, you'll easily be able to recognize when she's feeling under the weather.

▷ *Have a proper milking area.* By proper, I don't mean expensive, over-done, or Pinterest-worthy. I'm talking about an area that is clean and

A NOTE ON MILK PRODUCTION

The more you milk, the more milk your animal will produce! Most owners of dairy animals milk either once or twice per day (roughly twelve hours apart). Consistency is key. Show up at the same time every day to keep your animal calm, comfortable, and producing.

set up for milking. A small area in a barn, shop, or garage will do just fine. Preferably, it should be out of the weather so the ground stays dry. Wet ground equals mud. Mud equals dirty udder. Dirty udder equals no good. There is simply one rule: Keep it clean.

▷ *Have the proper equipment.* Again, I'm not talking about anything fancy-schmancy. But I am talking about common sense. Plastic buckets will not work. Plastic is porous, allowing the milk to get inside the pores. It's impossible to properly clean it.

▷ *Clean the equipment.* All of the basic milking equipment is inexpensive, very easy to maintain, and easily sterilized. It's not about having the most expensive or extravagant setup. Rather, milking safely is about doing a few things well. Buckets must be thoroughly washed, both inside and out (bottom too!) with hot, soapy water between each milking. The same goes for the milk storage jugs and the filter. It's all got to be really, really clean. All. The. Time. No slackin' on this.

Basic Milking Equipment

Two stainless-steel buckets, one for the milk and one for the wash water

Washcloths, washed between each use

One reusable coffee filter for filtering the milk

Glass jars for storing the milk

A good stainless-steel bucket is almost all that you need for milking.

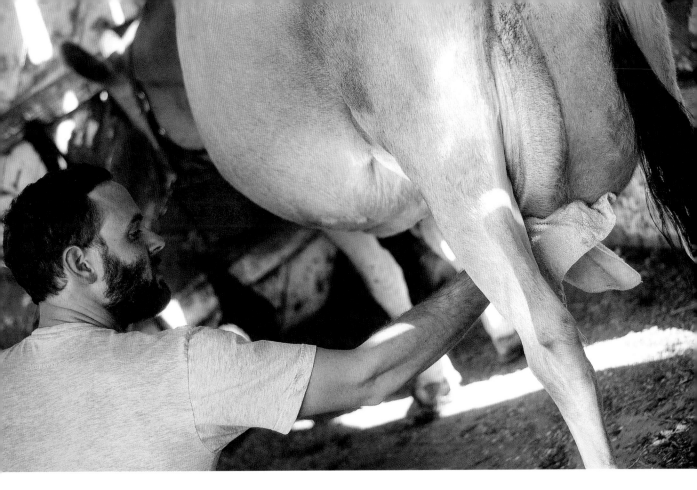

A clean animal makes for healthy, delicious milk.

▷ *Clean your animal.* Use a stainless-steel milk bucket to hold about a quart or two of hot, soapy water. Castile soap or a gentle dish soap works fine—usually just about a half teaspoon. At each milking time, the milker carries out one wash bucket with soapy water (and a clean wash rag) and an empty milk bucket. Take the time to gently wipe and scrub the udder, carefully removing any bits of hay or manure. The warm water helps to dissolve anything crusted on. The scrubbing is very effective at helping the goat or cow to let down her milk as well. If the animal has a particularly dirty udder that day, even after cleaning it properly, I'd still lean toward giving the milk to the animals. It's simply not worth risking any sort of contamination. You'll get plenty of milk; we don't need to be sticklers.

▷ *Milk cleanly.* Practice doesn't always make perfect, but you will get much better at milking cleanly. Remember to keep your hands clean,

MILKING MACHINES

The time that it takes to set up and clean a milking machine is about the same as it takes to hand-milk the animal. It goes without saying (but I'm going to say it anyway!) that the machine needs to be thoroughly cleaned between each milking. Break it down, clean it with hot, soapy water and allow it to air-dry, as towels can be transporters of unwanted bacteria.

too. A few pieces of hay or hair won't hurt much, but let's keep the poop at bay. After the animal is thoroughly stripped of all her milk, I like to spray the teats with a diluted iodine spray and leave it to dry. It keeps anything from getting in and infecting the teats while they're still loose and open.

▷ *Filter the milk.* You're going to get a few bits of things in there. That's okay. Just run the milk through a coffee filter to remove any particles from the milk.

▷ *Cool the milk quickly.* This is one of the most important steps in raw-milk handling. As soon as you've milked your animal, the bacteria in the milk begin eating and digesting the sugar in the milk. When drinking super-fresh milk, you'll notice how sweet it is. Each day after milking, it gets less and less sweet. Cooling the milk quickly helps to deter the growth of bacteria.

▷ *Handle with care.* Milk is a perishable product and should be treated as such. Even though raw milk doesn't go rancid like pasteurized milk (rather, it sours), it's still important to treat the product with extreme care. When raw milk is handled correctly, it's easy to feel comfortable not only drinking the milk but allowing our children, friends, and animals to drink it too. Contrary to what the FDA would have you believe, raw milk is not to be feared. It, like any other agricultural product, has the potential to be produced well and safely. Handled

To get the milk cool real quick, stick it in the freezer for an hour or so to get the cooling process rockin' before moving it to the refrigerator. Set a timer. You'll forget it's in there. Trust me.

Don't ever add warm milk to cold milk! Each animal's milkings should be kept in separate labeled, clean jars.

improperly, like any other agricultural product, it also has the potential to make consumers sick. It's all in the way we treat it and the respect we give the animals that provide it to us.

Raw Milk

I get asked quite often about our choice to drink raw milk. Not only do we drink it, but we've spent thousands of dollars and hours of energy acquiring it. Many wonder how we can feel safe drinking a product that is barely even sold for human consumption, illegal in many states, and sold as "for pet consumption only" in a dozen others. Is this product really that dangerous? So dangerous, in fact, that it requires homogenization, pasteurization, and the proper government label to make it safe? Surely not.

Raw milk contains more enzymes, nutrients, and beneficial bacteria than its pasteurized counterpart. The process of pasteurization was developed so that the milk could travel and be sold farther from the farm. In essence, it prolonged the milk's shelf life. Unfortunately, many delicate balances in the milk were destroyed by the pasteurization process. Lucky for you, since you'll be in charge of the milking on your backyard farm, you can control the quality of the entire process, drink the freshest milk you've ever tasted, and feel confident in the product you're consuming.

You can pasteurize your milk at home by pouring it into a double boiler and bringing it to 165 degrees F for at least 15 seconds, stirring constantly.

What to Do with the Calf

"But wait, what do I do with the baby?" Great question, observant reader. In the process of owning your dairy animal, you'll subsequently end up with little ones, be they calves, lambs, or kids. Many of these babies can be

A calf can make a wonderful milking companion . . . but you've gotta learn to share!

sold to other families, raised for meat, or raised as future milkers. Should you keep the little one, you have a few options with regard to the nursing and milking routine:

▷ *Leave the baby with the mom 24/7.* Pros of this method are that you don't have to worry about separate pens, shelters, waterers, or monitor any sort of interactions. Cons? The moms hold back the milk for the babies. I don't know how they do it. It's mothering magic.

▷ *Separate the baby every twelve hours.* Of course, this involves having separate space for the two and being able to keep them both in said separate spaces (easier said than done!), but it will allow you to get a full milking done in peace and quiet before bringing them back together. Most commonly, the little ones are separated at night, the mother is milked in the morning, and they are put back together for the day.

▷ *Bottle feed.* Yes, this involves milking the mother, putting milk into a bottle, and feeding the baby. Alternatively, some babies are fed milk replacement from the feed store instead of their mother's milk. Some farmers choose this method because it allows the baby to bond with you, which is ideal for future milkers, as well as making sure you get the "cream of the crop." Unfortunately, this means you're tied to bottle-feeding the baby until it's old enough to be weaned, which gets old quickly. Trust me on this one.

Once-a-Day Milking

Let's say that you've been milking your animal for a while now, and things are going great. You're getting all kinds of delicious milk to drink and play with in the kitchen. Unfortunately, that night milking session is putting a damper on your social life. Maybe you've got a local wine club to attend. And maybe you don't need *all* that milk anyway. Fear not! There's a great option available: once-a-day milking. Once-a-day milking still allows you to get fresh milk from your animal without having to milk as often. So now, you can attend the wine club without fear. Lucky you! Here's how to do it:

1. Pick the milking time you want to keep–morning or night.

2. Start by leaving about 25 percent of the animal's milk in the udder during the milking you want to eliminate. It's a little difficult to estimate since you can't see the milk in the udder, but once you've been milking for a while you have a pretty good feel for how much milk is left.

3. At the next milking, completely strip her out.

4. Repeat this process for three to four days, leaving some of the milk in during one milking and completely stripping during the other. Finish the cycle with a "milk in" day.

5. Now skip the milking you'd like to eliminate. But be sure to feed the cow as you normally do.

6. The next day, milk both times, leaving some milk in the udder during the milking you'd like to eliminate, as before.

7. Repeat steps 5 and 6 a few times.

8. Completely stop milking during the milking you'd like to eliminate. Feed as normal. Strip out at the other milking.

9. Watch carefully for any signs of chronic engorgement, which can easily lead to mastitis (a bacterial infection of the mammary gland). The key to moving an animal to once-a-day milking is to go slow. A gradual process is easiest and healthiest for her. Always watch for signs of redness or tenderness.

10. That's it!

What to Do with All That Milk?

Almost all dairy-animal owners hit that point where their refrigerator is, quite literally, spilling over with milk! Maybe you got behind in your cheesemaking, or maybe the animal is simply giving more than you expected. Regardless, you've gotta find a place for all that delicious milk to go! Even with four hungry little ones running around my feet, we still find ourselves with a surplus every so often. Need some ideas of what to do with it all? Here are some of my favorite recipes.

Homemade Ricotta Cheese

A true art form, cheesemaking is a great way to use up gallons and gallons of extra milk at a time. Unlike hard cheeses that can take years to master, soft cheeses like ricotta can be made in a very short period of time and are fairly easy, even for the novice cheesemaker.

4 cups raw milk

3 tablespoons freshly squeezed lemon juice

½ teaspoon sea salt

1. Heat the milk on the stove until it reaches 190 degrees F, stirring to prevent a scorched pan.

2. Remove the pan from the heat. Add the lemon juice. Gently and slowly stir one time.

3. Let the pan sit undisturbed for 5 to 10 minutes.

4. While the mixture sits, line a colander with a cheesecloth (folded in half for double thickness). Place the colander over a large bowl to catch the whey.

5. Pour the curdled milk over the cheesecloth and allow it to strain for 2 hours. Do you understand why they call it cheesecloth now?

6. Once the mixture has strained to your desired consistency (the longer you let it strain, the thicker it will become), mix in the sea salt, and bam! That's how you make ricotta! Store the cheese in a sealed container in the refrigerator. And what about all that beautiful whey that was drained off? Save that in the fridge, too! It can be added to your morning smoothie or oatmeal, used to lactoferment produce, or at the very least, be fed to your chickens, pigs, dogs, or cats!

Junket

Junket is cheese . . . in a way. But it's totally not cheese. Junket is flan in a way . . . but totally not flan. And junket is custard-esque . . . but not at all. Junket is in a category all its own.

> **1 quart raw, organic milk**
>
> **1 tablespoon active kefir or whey**
>
> **½ cup maple syrup or honey (or to taste)**
>
> **¼ teaspoon salt**
>
> **1 teaspoon ground cinnamon**
>
> **½ teaspoon ground nutmeg**
>
> **½ teaspoon cardamom**
>
> **2 drops liquid rennet or ¹⁄₁₆th of a rennet tablet**
>
> **¼ cup filtered water**

1. Warm your milk gently on the stove in a small saucepan until it reaches 90 degrees F and then turn off the heat.

2. Add the kefir into the milk, cover the pot, and allow it to incubate for 1 hour. This will allow the good bacteria time to grow and flourish. It's like microbial science right on your stovetop. I can't be the only one who thinks this is insanely cool.

3. After incubating the milk, mix in the maple syrup, salt, and spices. Stir to combine.

4. Dissolve the rennet in the filtered water and add this to the milk. Gently stir to combine.

5. Pour the junket into cups or serving dishes of your choice. I use coffee mugs. Because they're pretty, yo. And that's important. Individual serving dishes work best.

6. Let the junket set for 1 to 2 hours, or until firm. Junket can be kept in the refrigerator until you're ready to serve, but I enjoy it best at room temperature and right away!

Homemade Butter

Butter is much easier to make from cow's milk because it's not naturally homogenized like goat's milk. This makes it easy to skim the cream off the top of the milk and turn it into butter! Yes, butter comes from cream. Essentially, butter is fat separated out of the milk. Hello! That's why it's so dang delicious! Fat is flavor, baby. When I've got a refrigerator that's chock full of milk, I often make up all the butter I can, freeze the extra, and give the newly skimmed milk to the various critters around the farm.

Fresh cream

1 tablespoon buttermilk (optional, for culturing)

1. Combine the cream and buttermilk together in a bowl. Let sit at room temperature for 8 hours. This is the optional "culturing" part of cultured butter and is a process of fermentation. During the culturing, bacteria convert the milk sugars into lactic acid. The result is a much more flavorful butter. It makes it, well, more "buttery." If you're omitting this step, simply start at step 2.

2. Pour the cream into a stand mixer. (Alternatively, you could pour the cream into a butter churn, but using a mixer will save your arms. Trust me.) Turn it on medium speed and allow the cream to whip. It'll start to look just like whipped cream, which is what we're looking for. But then you want to keep going! Here are the stages you can expect during the mixing: Cream. Whipped cream. Thick whipped cream. Lumpy whipped cream that will start to fly all over your counter (cover your mixer with a towel to keep this from happening). Weird, funky, chunky-looking whipped cream. Liquid with bigger chunks that no longer resemble whipped cream. Large chunks of butter floating in milky liquid.

3. Remove the butter chunks from the bowl and knead together to combine. Run under cold water and knead the butter for three minutes, or until buttermilk is no longer running out of the butter. You'll notice that the buttermilk has a milk-like color. It's important to get as much of the buttermilk out of the butter as possible, as this will cause it to spoil quickly. Often, I'll stick my entire slab of butter in a bowl of cold water and just massage it gently. Then, I'll dump the water out, refill the bowl with fresh water, and continue to massage until the water stays clear.

4. At this point, the butter can be eaten, salted, or frozen. I simply wrap my butter ball up in a small piece of parchment and tuck it into a bag in the freezer for preservation. I'm always stocking up! Girlfriend can't ever have enough butter stored up.

FEED EXTRA MILK TO YOUR ANIMALS

Homegrown pork is all the more delicious from pigs that have been fed milk! They're a great way to use up large amounts of extra milk and—ta da!—grow more bacon from it. Pigs, laying hens, and meat chickens can all make great use of that extra milk. They don't even mind if it's soured. Soak their grains in the soured milk for a few days and they'll love you all the more.

Homemade Yogurt

Why the heck not turn some of your milk into yogurt and enjoy it with some fresh berries from the garden and a drizzle of honey from the bees? See? I told you this was the good life.

1 gallon fresh milk

1 cup of plain, organic yogurt

1. Pour the milk into a large saucepan. Heat the milk slowly over medium heat, stirring occasionally, until it reaches 180 degrees F. Use a thermometer for this. Turn off the heat under the milk and allow the pan to sit until the milk reaches 110 degrees F (30 to 60 minutes).

2. While the milk is cooling, preheat the oven to 200 degrees.

3. Add in the yogurt and whisk to combine the milk and yogurt. This yogurt will give us the active, live cultures we need to turn the milk into yogurt! Isn't that cool? I heart bacteria.

4. Pour the yogurt mixture into glass jars with lids. Wrap each of the jars in a tea towel to help keep it warmer longer, and place them in the preheated, but now turned off, oven. Shut the oven off and allow the jars to culture in that warm oven for 6 to 8 hours.

5. Remove the jars, place them in the refrigerator, and allow them to cool. The yogurt will continue to thicken as it sits.

6. Sweeten with honey or maple syrup, to taste, if desired. Umm, delicious!

Kefir

Kefir is rich in probiotics and easy to make!

Essentially, kefir is fermented milk. "Grains" are introduced to milk, which ferment the milk and culture it to a rich, tart liquid. It's fermented, and therefore, has a slightly sharp flavor to it. Think about kefir as yogurt on steroids. Millions and millions of beneficial bacteria, just swimmin' around in that milk for your digestive pleasure.

Kefir grains

Fresh, unpasteurized milk

1. Acquire milk kefir grains (about a teaspoon will get you started). These grains are a culture of bacteria and yeasts that break down the milk and make nutrients more accessible to our bodies. They look like rubber cement boogers (What? I couldn't have been the only one that made those in elementary school!).

2. Fill a mason jar with milk. Add the milk kefir grains.

3. Set the jar on the counter and leave it out at room temperature to culture for 24 hours.

4. At this point, the entire jar can be put into the refrigerator until you're ready to drink the kefir. When you're ready, simply dump the contents of the jar through a fine mesh strainer. Use the back of a spoon to gently press the cultured kefir through the strainer, while reserving the grains.

5. The liquid kefir can be utilized in a variety of ways, such as in smoothies. The grains will go into the next jar of milk for culturing! And so the cycle continues.

Neighboring

Awhile back, my husband and I had the privilege of visiting some of our friends in Wyoming. After finding someone to cover the farm for those few days, as well as a set of welcoming arms for the three children we were leaving behind, we made the trek via planes, trains, and automobiles to arrive on their doorstep. While the wife and I sat inside discussing farms and kids and business and life, the menfolk stood out in the shop and discussed . . . I don't know . . . man stuff. Another gentleman, who was their closest neighbor from more than six miles away, came over to stand around and probably drink a beer or two while talking about said man stuff. You know what he called it? "Neighboring."

Truck run out of diesel? Building a barn? Time to harvest the pigs? Fence get knocked over? Got a few extra servings of chicken soup for

'Round these parts, neighbors lend each other livestock trailers and help to trim sheep hooves.

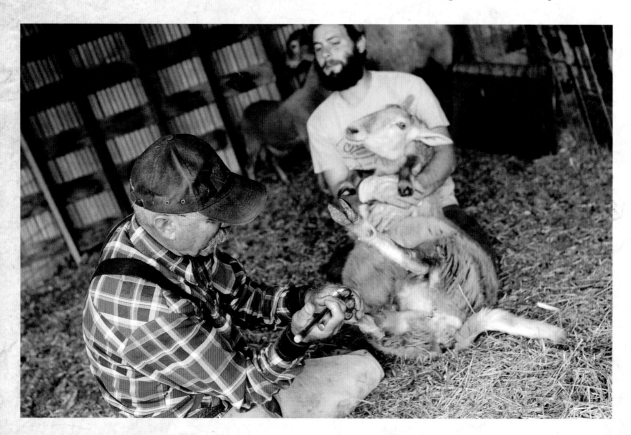

supper? Perhaps an extra loaf of bread to share? Or maybe just in the mood to chat about the weather? That's where "neighboring" comes into play. Because the reality is this: There is an entire community of people who know the struggles and triumphs of this lifestyle. These are the people in your community who know what it means to help and seek help in times of desperation, who know how to barter so everyone can get what they need, and who know how to celebrate in times of good and plenty.

As much as we pride ourselves on the independent lifestyle we've come to love and appreciate, the fact is it's not meant to be done alone. And frankly, it's not best to do it alone either. Many of the farm's wonderful blessings are only amplified when we get to experience this life alongside others—when we're "neighboring."

When we moved to our now-farm from our old-farm, we did it in the dead of the worst winter our valley had seen in over a decade. Our ten tons of hay were covered with tarps that were hidden under four feet of snow. The driveway was so icy we had to park at the bottom and walk up the steep grade to the house with every single box we were unloading. It was so cold that any venture outside required heavy coveralls, scarfs, hats, mittens, and thickly lined boots. Our family began to resemble a tribe of marshmallow people awkwardly waddling around the land.

When it felt as if the world was surely going to cave in, my gigantically pregnant self cried for help. And my community heard me. My parents showed up with trailers, baskets of food, and strong backs. Others helped to take loads to the dump and pack up boxes. Friends volunteered to watch children, move our livestock in their trailers, and bring us wonderful meals to ease the burden. My dad spent many days ripping up and hauling old shag carpet from the new cottage and my brother-in-law came almost immediately to start helping us build our garden and overhaul the yard. My niece showed up when it was time to butcher our pigs to help us scrape and scald. Our community surrounded us at a time of need.

Luckily for us, "neighboring" isn't only reserved for times of need. It also includes sharing in the joy of harvests. Extra eggplant? Take a basket to your friend. Plumber coming by to fix a leaking pipe? Send him home

with a dozen farm-fresh eggs. Scored a few bushels of apples? Invite a
family over for some cider pressing! Sharing your excess with others and
blessing them is truly the "culture" of the farmer. One of my very favorite
ways to "neighbor" is to feed people. I love nothing more than to prepare a
special feast. We bring out the appetizers–perhaps marinated homegrown
tomatoes on top of homemade toast–and, of course, the good wine. We'll
sit around nibbling and fellowshipping over some homegrown pulled
pork sandwiches and caramelized fruit. We'll walk around the farm and
talk about the new additions to the barnyard, who's due to lamb soon,
and give all the animals a good once-over. We'll talk about dreams, plans,
and struggles. We'll share not only the pains of the farm life, but also the
immense pleasures it can bring.

Neighboring involves welcoming someone onto your farm and, in a
sense, into your family. It's trading honey for use of the neighbor's bull to
breed the family cow or swapping extra Swiss chard for garlic. It's covering
for each other in times of need and caring for their farm as you'd care for
yours. Neighboring is a way of life that naturally comes with the territory
for the farmer–we've all chosen a way of life that is out of the ordinary
and off the beaten trail. And we're traveling this road together!

CHAPTER 5

Raising Meat in Small Places

While growing meat might not be the first thing on your list as a backyard homesteader, it should most certainly make its way there as your farmgirl experience and confidence grow! Growing your own meat can be a huge benefit to your family. Not only will you get tastier and healthier meat, but you'll gain an appreciation for where that meat comes from. Before you begin, it's good to ask yourself a couple of questions:

1. What meat do you like to eat? Ain't no use growing it if you don't want to eat it.

2. How much space can you devote to raising meat? This will largely dictate what type of animals you raise. I'll cover the easiest and smallest of the meat animals, like chickens and rabbits, before venturing into larger animals, such as pigs and sheep.

Broilers

We run our meat chickens in a very large, open, outdoor pen where they have access to grass, bugs, weeds, and sunshine. Even growing enough chickens for our family to last throughout the year, we still only use a small fraction of land for their production. They're usually harvested before they reach thirteen weeks of age, so the land and time commitment to raising them is minimal. The meat from home-raised birds is *fantastic*, and one can easily become spoiled by the vast taste difference from commercially raised chicken.

Expected Harvest

Depending on what breed you raise, you can estimate that your harvest from each bird will be between four and eight pounds. We raise Freedom Rangers on our farm and they have faithfully dressed out (that is the weight after all the feathers and guts have been removed) at around six pounds. Feed and genetics will greatly affect the expected harvest per bird. The commercial Cornish X will likely dress out at around seven pounds and in a shorter timeframe, however, many would argue, at the cost of flavor! Broilers that free-range will grow much slower than those that are raised on a broiler grain, which is high in fat and protein. Most formulated feeds contain corn, which is popular for putting fat on a bird faster than a standard grain. Depending on genetics and diet, the birds could potentially be ready for harvest in as little as seven weeks (though the slower-growing heritage varieties are usually ready around twelve weeks). I don't mind investing a bit more time for a tastier bird.

Rain, snow, or shine, it's chicken-harvesting time.

Requirements

Broilers require a small area to roam and grow. While they'll take as much room as you can spare, just like a laying hen, it is possible to keep the birds in a small run while they're growing. The birds can be raised in small batches if you've only got a small amount of room to work with or in a large group if you've got the space. It goes without saying that the birds should always have access to shelter and protection from the weather, free range of food, and fresh water. Predators are almost always a threat to chickens (the poor, defenseless chicken!) so make sure you've got some protection in place. Let's pretend I haven't lost thirty birds in one night to an owl. Some memories are best forgotten (stupid owls)!

Shelter

If you'd like to set up a temporary shelter for your chickens, a few metal T-posts tapped into the ground and chicken wire will do just fine! Fine mesh netting over the top will keep the birds safe from aerial predators. If you've got a crate or small coop for the meat birds, it's a great idea to shut them in at night to help protect them from raccoons, coyotes, and even wandering dogs. Electric poultry netting is also a great option if you'll be grazing them out in the open and want to move them around your property as they grow.

Butchering Basics

Butchering is never fun, but it's a necessity for those of us who have taken on the joy, and burden, of raising our own meat. I won't pretend that I don't often pawn off the actual killing on my husband. Chickens are a great starter animal for you to learn basic butchering skills. Prepare yourself for feathers–lots of them. And make sure you've got a sharp knife!

How to Butcher a Chicken

1. Place the chicken upside down in a kill cone. Using a sharp knife, slit the artery in the throat (which runs right on the backside of the

Butchering isn't fun, but it's a necessity if you raise your own meat.

An empty, gallon-sized plastic jug can easily be made into a kill cone.

earlobe) and allow the blood to drain out and the chicken to die. This usually takes around 30 seconds to 1 minute. Apply more pressure to the knife than you think you'll need. Getting a knife through the feathers can be tough and multiple attempts are not desirable for anyone involved, chicken or human.

2. Once the chicken is dead, remove its body from the kill cone.

3. Gently dip the chicken into a large pot of 145- to 150-degree F water for 3 seconds, shaking it gently while it's submerged. Pull the chicken from the water for 3 seconds. Dunk it again in the hot water for 3

seconds, shaking it gently. Again, pull the chicken out of the water for a few seconds. Grab a feather from the bird and pull it out. Does it slip out easily without resistance? If yes, proceed to the next step. If not, continue to dunk the bird for 3 seconds at a time until the feathers pull out like warm butter.

After scalding, the chickens can easily be hand-plucked or plucked utilizing a chicken plucker (which are available for sale from online retailers).

4. Once the chicken has been scalded, begin plucking the feathers by hand or transfer the bird to a plucker. To pluck with a machine, just place the scalded chicken into the plucker, flip on the switch, and spray the chicken with water while the machine is running (this helps to remove all those little feathers that would otherwise stick to the skin). This makes it much faster and much easier!

5. Transfer the chicken to the "evisceration station" (as we like to call it). At this station, you will remove the head, feet, and guts.

6. Halfway up the chicken's legs is a joint that connects the leg to the foot. Cut right in the middle of the joint so the feet can easily be removed. I use a filet knife for this and it works just fine.

Don't throw those chicken feet away! They look a bit creepy but make a wonderful addition to stocks.

7. Next, the head. For this you may need a pair of regular garden pruners. It can be a bit tricky to remove. Once the neck bone is broken, though, it's easy enough to cut through the remaining skin. Bust out your meat cleaver if you'd like.

8. After removing the feet and head, flip the chicken over so it's breast-side up. At the narrow end of the breast, grab the skin with one hand and use a knife to cut a slit in the skin. Get all your fingers in there and use your hands to gently pull open the body cavity, revealing the innards.

Ducks, turkeys, geese, and quail can easily be raised on your homestead. The same requirements and butchering methods apply.

9. Once you can see the innards, it's easy to reach in with one hand and pull out all of the guts. Using your fingers, it's nearly impossible to rupture the intestines, but be careful nonetheless. It's important that no poop gets onto the meat, as this will contaminate it. Pull, pull, pull. At this point, you'll likely notice that there is a long straw-like thing at the top of the bird that's keeping you from removing the guts completely. That's the windpipe. Reach in there, wrap your fingers around it, and pull really hard. It'll come out. There should be nothing left in the body cavity when you're finished.

10. But, wait! The guts are still attached to the chicken, aren't they? Yes, they are. So what you have to do is take a sharp filet knife and gently cut a V shape around the vent of the chicken. *Be careful* not to puncture the intestines or the vent. Just cut around it and remove it completely.

11. Place the chicken into a large bucket filled with ice and let it remain there while you finish harvesting the rest of the birds if you're doing more than one.

12. Once all the birds are cleaned, I take them inside to wash them out and do quality control. Any stray parts left in the body cavity are removed, leftover feathers are plucked, the entire chicken is rinsed with water (both inside and out), and all of the chickens are placed into individual shrink-wrap bags, along with the head, the feet, and any other bits we will keep.

13. Place the chickens, in their open shrink-wrap bags, into the refrigerator for one to two days. This will allow the chickens to "air-chill"–an incredibly important step in the process. Without this, the chicken tends to be stiff and chewy. During the air-chilling, the meat has a chance to rest and relax, resulting in a much more tender bird. After this time, shrink-wrap the birds (following bag manufacturer's instructions) and move to the freezer. Booya.

Shrink-wrap bags keep the chickens safe from freezer burn.

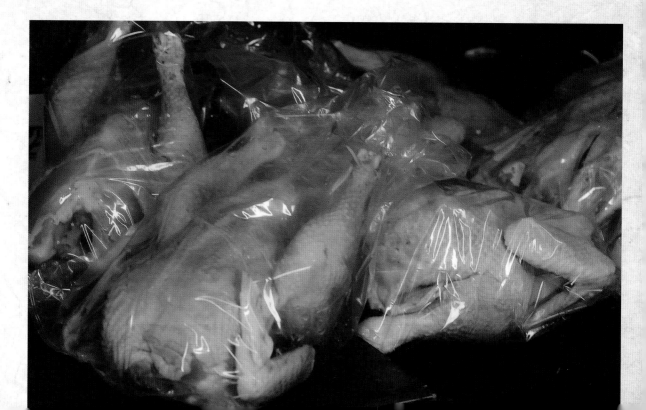

How to Cut Up a Chicken

"Piecing out" a bird is a skill most of us have lost in this day and age, thanks to the supermarket. It's funny, but here on the farm, meat doesn't come in packages. Whole, skinless, boneless, chicken breasts? Nope. As convenient as that may be, it just ain't the way the good Lord designed things. Chickens come with two wings, two thighs, two drumsticks, and two breasts. The neck, feet, and carcass can be used for stock, and the liver is fabulous mixed up into pâté.

I've gotten a lot better at piecing out a chicken over the years, though I'm still far from an expert. But that's never stopped me from anything. Here's how you do it:

1. If the chicken still has its feet attached, you can simply cut those off at the joint by gently pushing through the joint with your knife. They'll come off very easily. Save the beautiful feet for the richest chicken stock ever. Same goes for the neck.

2. I like to start with the breast. Lay the bird on its back with the breasts facing upward, toward you. You'll know the breasts when you see them because, well, they're breasts. The back of the bird has a defined spine that runs down it and is stiff to the touch. The breast side of the bird is squishy and smooth.

3. Find the center line that runs between the two breasts with your knife and gently begin to fillet the breast. A bone runs right through the two breasts, so your knife can simply stick close to the breast bone as it fillets the breast away from the bird.

4. With the chicken still on its back, it's mighty easy to push the legs down to the counter with your palms, which will separate them a bit from the carcass. Then, all you need to do is simply use your knife to gently cut through the skin and thin pieces of flesh that attach the leg and the thigh to the bird. You'll continue this as you also remove the thigh from the bird, separating it at the ball joint where the entire leg attaches to the side of the chicken.

THE BEST ROAST CHICKEN

The very best chicken is made by simply covering an entire broiler chicken with about 4 tablespoons of olive oil before sprinkling liberally with fresh or dried herbs of choice. Rosemary, lemon thyme, and parsley are my personal favorites. Roast, breast side down, in a 375 degree F oven for 2½ to 3 hours until the skin is crispy and the juices run clear. Let the chicken cool slightly before fighting off your various family members for the drumsticks. And Amen.

5. Lastly, remove the two wings by using your knife to push through the joint. Don't try to cut through the bone, just find that joint and cut through the cartilage and connective tissue—this is much easier.

6. Last, if you're super talented, you can remove the "oysters." These are two teeny tiny little pockets of the most flavorful and tender meat on the entire bird, located on the bird's back right above its thighs, on either side.

7. Ta da! I knew you could do it.

Sheep

As much as I love a good roast chicken, I'll confess that lamb is my favorite meat that we grow on our teeny tiny farm. We keep a small herd of Katahdin sheep, with just a small handful that lamb each spring. Typically our ewes give birth to twins, which means that we're left with quite a few lambs to butcher for meat. These lambs are butchered each year right on our farm and are the most anxiously anticipated meat in our freezer each fall. Though it's obviously not as common as beef, lamb holds its own when it comes to sustainability and flavor. Many people who say they don't like lamb have never really eaten meat from a quality animal, prepared well. For this farmgirl, however, growing lamb is an incredibly rich and beneficial experience.

Basic Requirements

When I was learning to raise sheep, I was told that sheep are the only animals that are "constantly trying to die." We've lost lambs to neighborhood dogs, escapades into the chicken coop where they consumed too much grain, bad mothers, and even to poachers. Girlfriend is just trying to get some meat in the freezer! That being said, our sheep bring great joy to my life (have I said that about every animal thus far?). Nine months of the year, I watch our sheep graze from my bedroom window. Heads to the ground, they happily nibble at the pasture grass before sunning themselves and frolicking around. Yes, sheep actually frolic, and it's wonderful to watch.

Sheep bring delicious meat to your supper table.

Sheep require little more than grass, water, and shelter. They love to graze when the season and grass allow it, but if they've got good genes, they can hold their condition well on just grass or alfalfa hay. Because sheep are so sensitive to bloating, we choose to raise our lambs exclusively on grass, omitting any sort of grain whatsoever. Fresh water with a bit of apple cider vinegar keeps them hydrated and worm-free. Many sheep farmers choose to dock (crop or remove) their sheeps' tails to help prevent fly strike, which is a nasty condition caused by flies laying eggs in the anus of the sheep. A docked tail helps to keep the hind end of the sheep free from trapped debris and fecal matter. Because our sheep are a breed that doesn't grow fleece, I don't bother docking their tails. During the height of the summer when flies are at their worst, I find it's easy enough to do a butt check every few days to watch for any problems. What, you don't butt check your animals?

When you're not busy checking their backsides, spend a bit of time making sure they've got shelter that will protect them from harsh weather. Give 'em shade and grass to lie in during the summer. Give 'em protection from snow and warm bedding in the winter. Sheep tend to be a bit flighty, meaning they're quick to run away from danger, loud noises, or anything that otherwise startles them. This is good to take into consideration when you're building fences. If they can run through it, squeeze through it, bust through it, or break it down, they will.

Sample Annual Sheep Calendar

Breed in October

Lamb in March (145 days later)

Wean lambs in July

Rebreed again in October

Butcher lambs in December

Basic Sheep Terminology

Ewe. A female sheep that will be bred and raise lambs.

Ewe lamb. A female lamb. Ewe lambs reach puberty anywhere from five to twelve months of age.

Lamb. A male or female sheep that is less than one year old.

Lambing. At some point in the birthing process, the ewe may need a helping hand from you to successfully deliver her lamb.

Ram. An uncastrated male that is used for breeding females. Typically, on a small farm, one ram is all that is needed. He can service up to twenty-five ewes during the breeding season.

Ram lamb. A male lamb. Ram lambs reach puberty at roughly five to seven months of age.

Weaning. The process of removing a nursing lamb from its mother. It is most cost-effective to feed the lamb directly rather than feed the ewe

Ewes, like our sweet Noel, remain on the farm and birth lambs each spring.

to produce milk to feed the lambs. That being said, as a backyard home-steader where profits aren't necessarily our goal, we have the freedom to allow the lambs to nurse for a bit longer. Mama's milk is so beneficial for them and makes the meat oh-so-tender! Early weaning can be done around sixty days. Because most backyard farms only have so many pens for animals, it's certainly possible to allow the lambs to nurse until Mama kicks them off. And trust me, she will when she's ready.

Wether. A castrated male. Castration is typically performed within a few days of birth and will prevent the ram lamb from being able to breed females. This is typically done for ram lambs that will be raised in the same vicinity as fertile ewes.

Lambing 101

What's that, you say? You've never helped to deliver lambs before and wouldn't have a clue what to do if there was a problem and you're feeling incredibly unprepared because the lambs could come, like, now, and you're totally not ready for an emergency? I've been there. Let me help.

Give Mama a few minutes to get acquainted before you swoop in to take a peek!

1. Let's start by preparing a dry, contained, warm place for the ewe to have her lamb.

2. If any problems arise, it will help if she's contained. Plus, this will make it easier to keep an eye on her and the lambs.

3. Now, let's make sure there's fresh water and hay available for her to keep her comfortable.

4. Just in case, let's also put together a little lambing box (see sidebar).

After the lamb arrives, spend a bit of time getting it used to you and your presence. Make sure it's standing and nursing from Mama. If you have any questions or concerns, call your veterinarian! Paying for a farm call is always better than a sick or dead animal. We usually allow our lambs to remain in a safe pen with their mama until we're positive they're eating and strong enough to stand up to a bit of push and shove from life in the herd.

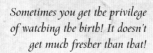

Sometimes you get the privilege of watching the birth! It doesn't get much fresher than that!

Butchering

Lambs are typically butchered when they weigh around 125 pounds, which, depending on genetics and diet, is usually around ten months of age. The dressing percentage (the amount of the live animal that ends up as a carcass) is 50 percent. From a 125-pound live animal, you'll end up with about 60 pounds of meat. Because we're producing this meat in our backyards, we can ensure not an ounce is wasted. Use the blood as fertilizer; compost the intestines; feed the kidneys, heart, and various bits to the chickens or dogs; livers should be eaten by *you*; and the head can be boiled for head cheese. Keep the bones for rich stock.

Basic Method

1. Separate the lamb from the rest of the herd. Don't allow it to feed for twelve to twenty-four hours. This will ensure the rumen (part of a sheep's stomach) is empty come evisceration time (an empty gut is much easier to work with than a full one).

2. Kill the lamb. Some prefer to slit the lamb's throat. Others choose to shoot the lamb. I would only recommend this method to a seasoned marksman. A sheep's head, particularly a ram's, is designed for impact and is notoriously challenging to penetrate correctly with a bullet. It's essential that the animal is calm and that the kill is instantaneous. I hate this part every time. Every single time.

3. Hang the lamb. A strong tendon on the back of the lamb's ankle (do sheep have ankles?) can easily suspend the weight of the animal. We typically hang our lambs from a tree limb, though a barn or garage rafter would be a great option as well. This will suspend the animal, making it easier to skin and eviscerate.

4. Skin the lamb. There are a million methods to skinning a lamb, but the basics are very simple. Cut around the top of the leg and use a sharp knife to gently separate the skin from the flesh. From the legs, cut inwards, along the butt of the lamb, toward the anus, which should be tied off with a string to ensure that nothing leaves the backside

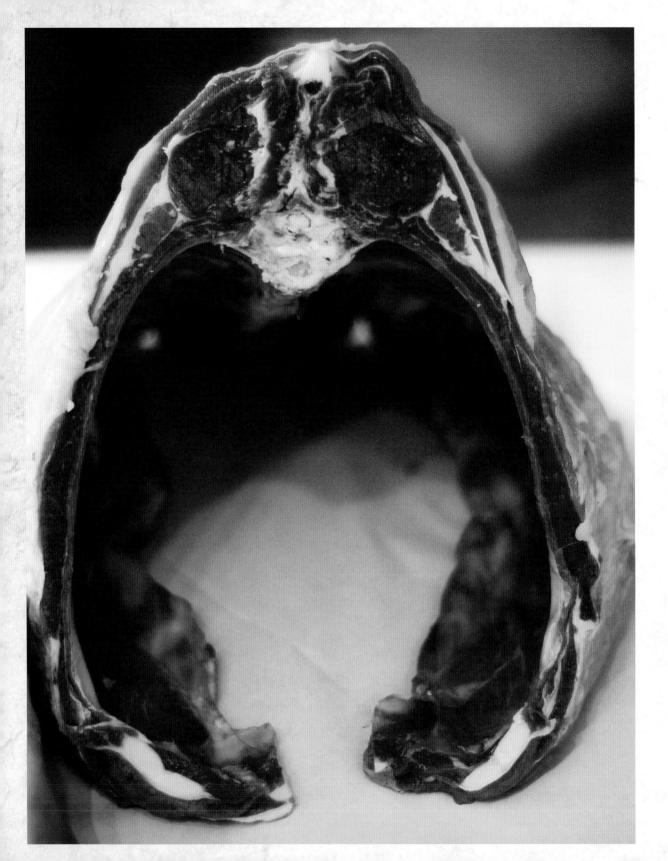

and contaminates the meat. After you cut around the anus, it's easy enough to pull the hide off (sort of like pulling a sweater off) until you get to the head.

5. Remove the head. Gently cut through the flesh at the top of the neck until you hit the spine. Located a vertebra and cut between it and the next one. Use your hands to twist the lamb's head until it pops off, using your knife to separate any spare bits of flesh. You should now have a skinned, headless carcass. You can remove the front hooves at this point, if you wish, by locating the first joint, bending the hoof with your hand, and using your knife to separate it at the joint.

6. Eviscerate the animal. The process is very similar for every animal. *Do not puncture gut-town, man.* A small slit at the top of the hanging animal will get it started. Continue cutting down the belly of the carcass with just the tip of your knife, making very shallow cuts. Using your hands, work the innards free. I prefer using my hands because I'm much less likely to puncture the guts with my fingers than with a knife. All that's in the body cavity needs to come out, by any means necessary.

Killing the animal is hard. Expect it to be.

7. Rinse the lamb. Your lamb should be hideless, headless, hoofless, and gutless. It should resemble a carcass hanging in a butcher shop. One final rinse of water will ensure any stray hairs or dirt have been removed.

8. Age the lamb. We've been known to cover ours in a canvas bag and hang it in an outside tree for one week to age. The temperature should be between 36 and 43 degrees F at all times. If you butcher during a

The cuts of lamb should be thoroughly wrapped in plastic and butcher paper to protect them in the freezer.

warmer time of year, you'll want to refrigerate the uncovered carcass for one week. This helps the flavors to develop, the muscles to relax, and some of the moisture to dissipate.

9. Butcher the lamb into cuts of choice. Wrap the cuts in a generous amount of plastic wrap and then butcher paper before labeling and storing them in the freezer.

Rabbits

Oh, am I ever a sucker for our meat rabbits. If any animal was too cute to eat, it'd probably be the bunny. Not that that actually stops me, but at least I'm acknowledging their furry sweetness. We've been raising our Champagne D'Argent rabbits for a few years now and every time we get a litter of them in the nesting box it feels like a miracle. And it is! A new litter can appear as little as thirty days apart from the last–wowza! That's some serious gestational power.

Why Meat Rabbits Are Totally Awesome

They fatten on grass. Grass, people! They're sustainable. Easy to grow. Easy to store. Inexpensive. Most of the year they can just be fed on weeds you pull from your garden. No grain. No alfalfa pellets. No money!

New kits will remain in the nesting box for a few weeks while they gain strength, grow fur, and wait for their eyes to open.

They produce rapidly! Rabbits have around a thirty-one-day gestation, which is incredible considering they can produce anywhere from one to fifteen (or more!) kits from each pregnancy.

The meat is wonderful. Rabbit meat is incredibly lean and is wonderful roasted, baked, fried, braised, and boiled. Because rabbit is so lean, it does well with some fat added in (bacon, anyone?). Don't tell me you don't like rabbit. Just don't. Because chances are you haven't had it. And if you have, then it's time to try it again. Because while it doesn't taste exactly like chicken, it is a very comparable meat and is much more sustainable for the small homestead.

Common Meat Rabbit Breeds

- *Champagne D'Argent.* I put Champagnes on the top of the list because I love them the most. And since I'm writing the list, well, that's what you get. Champagnes have been used for meat for centuries and grow up to ten pounds. They have silver fur with shades of black along their faces and paws. They're also super sweet, for what that's worth.

- *California.* A cross between the New Zealand Whites and Chinchillas, the California rabbit is large and grows quickly, typically weighing in at eight to ten pounds.

- *New Zealand Whites.* A large breed that grows quickly and is one of the most common types of meat rabbits available. They have white fur, pink eyes, and typically weigh in at ten to twelve pounds.

- *Chinchilla.* The chinchilla rabbit resembles a chinchilla, but is much larger, and can grow up to nine pounds. They are raised for both their meat and their hides, which are a beautiful grayish-tan color.

- *Silver Fox.* Much like the name suggests, Silver Fox rabbits are black with silver shading. They average around nine pounds and are considered a "fancy" rabbit. *Ooh la la.*

- *Rex.* Another common meat rabbit that's also raised for its spotted hides, Rex rabbits come in a huge variety of shades and colors and typically weigh ten pounds.

Rabbit Shelters

In nature, rabbits will dig deep warrens into hillsides, which provide them steady temperatures through the weather. Because they're designed this way, it's important to keep in mind when raising rabbits. They need warm bedding and protection from rain and snow in the wintertime. Rabbits are sensitive to heat and often require a

The Rabbit Cast of Characters

Buck: A male rabbit used for breeding purposes

Doe: A female rabbit of breeding age

Grower: Rabbits that will be raised and harvested for their meat

Kit: A baby! Squee!

Kindling: When a doe is giving birth to a litter

Grower rabbits can be kept in a basic enclosure.

bit of TLC during the height of summer. Frozen water bottles, misters, and shade will serve them well. As you set up your rabbitry, keep this in mind. They will need to be placed in a safe, secure, and protected area.

In addition, does will need nesting boxes. They will build nests in these to protect their kits after birth. For the first week of life, the baby bunnies are blind, furless, and completely defenseless. The doe will pull lots of her own hair to build a warm little area to give birth in, but it's very common for the little kits to flop out of the nesting box, so try to keep it as secure as possible. Always provide ample bedding, such as straw hay, so the doe has something else to pull from as she builds her nest. When a litter is born, it's a good idea to keep an eye on them every couple of hours to make sure no escapees have found their way out of the box. They'll need the warmth of the nest to survive. It's not uncommon for the doe to eat the weak, disabled, or runts in the litter. I know. Gross. But now you've been warned.

Keeping Them Safe from Predators

A rabbit's natural defense is its speed and ability to hide in the warren. Without this (such as when they're raised in a cage), they're incredibly helpless. Give the poor things a fighting chance by safely securing them far and away from raccoons and dogs. If you raise your rabbits out in the open, keep in mind that they're a perfect-sized snack for an owl or hawk. See what kind of lessons I've learned so you don't have to?

Feeding Rabbits

Rabbits do very well on timothy grass hay and a variety of grasses, leaves, and weeds from around your property. Clover, dandelions, and sunflower seeds are an absolute favorite treat, as are bananas (who'd a thought?). Many homesteaders feed their rabbits pellets, which allows for easy feeding and minimal effort, though they're slightly more expensive. For the most part, rabbits should be fed "free choice," meaning they can have as much as they'd like all the time! Breeding does should always be given

Rabbit manure is great compost for your garden! When I clean out the rabbit's hutch, I sprinkle it on my gardens. Happy flowers!

free-choice feed as their bodies work to grow more babies and lactate. They often thrive from a small handful of rolled oats as well during peak lactation. If you learn one thing about animals from this book, it's this: Take care of those lactating mamas!

Rabbits can be fattened on grass and garden waste.

Harvesting Rabbits

Two breeding does and one buck can provide your farm with up to fifty rabbits a year to eat. Dressed out at about 3 pounds each, that's about 150 pounds of meat! The rabbits will nurse from their mother for five to six weeks before they can be weaned and fed in a different area. They will continue to grow for another six to eight weeks until they weigh in at about six pounds. Rabbits can be dispatched easily with a pellet gun and can be skinned and harvested in the same manner as a lamb. The pelts can be cured and utilized.

Pigs

I know what you're thinking. *Pigs?!* Come on, Shaye, I have some *pride* in how my backyard smells. Dear friend, open your mind and your heart. Kept in open spaces and fresh air, unlike commercially grown pigs, the pig is an animal of pure wonder. In fact, our pigs are the cleanest members of our farm! They dutifully poop and urinate in only one small corner of their pen (I'd like to think it's because they love and appreciate me). We've been raising pigs for years now, and I'm amazed at how beneficial they are. Each morning, after clearing the breakfast table, I scrape the half-eaten eggs and pour the leftover glasses of milk into the scrap bucket that sits on my countertop. Throughout the day, I'll throw in banana peels, apple cores, eggshells, coffee grounds, stale bits of bread, and old oatmeal. Any and all edible scraps get put into that bucket. The following morning, that scrap bucket will be fed to our pigs, along with a small bucket of fermented grains. And you know what? Those pigs, God bless 'em, will take those nasty scraps and turn them into *bacon*. Pigs are incredibly efficient at converting waste into meat. Even just one pig a year will serve your family with plenty of pork to enjoy. I'm going to need you to put your assumptions aside about what it must be like to raise a pig. Keep reading!

Meat Pigs and Breeding Pigs

As with most animals, pigs tend to fall into two categories: breeding pigs and feeder pigs. Breeding pigs are ones that will remain on your property for years to come and whose sole purpose is to breed and create new piglets! The average backyard homesteader looking to raise pork for her family won't dabble too much in breeding pigs. Rather, she'll tend to pick up a feeder pig each year at the market or a local farm. Feeder pigs are raised exclusively for meat. They are typically purchased in the early spring after being weaned from their mother and will continue to grow and fatten through the summer and fall before being butchered the following winter.

> ## The Pig Cast of Characters
>
> **Barrow or hog:** A castrated male pig, typically raised for meat consumption
>
> **Boar:** An intact male over six months old that is kept for breeding purposes
>
> **Gilt:** A female pig that has not yet borne a litter
>
> **Piglet:** A baby pig
>
> **Sow:** A female pig that has given birth to a litter

We purchased our first little weaner pigs from "Farmer Steve" down the road. While driving down his gravel driveway, we'd have to brake quickly for litters of pigs that would quickly dart across our path. That's the moment I first fell in love with pigs. Prior to Farmer Steve's pastured pigs, my only experience had been with commercial operations, which left a sour taste in my mouth and an intensely disgusting odor on my clothes. Walking through concrete bunkers with pigs in crates not even big enough to turn around in, I couldn't quite fathom how we'd gotten to this point in agriculture. These pigs were far from experiencing a happy and healthy life prior to their slaughtering.

It may have been a small moment in the history of the world when we purchased our first pig from Farmer Steve, but it was monumental in our personal farming journey to take a stand against commercial pork production. We now have a few breeding sows and a boar on our small plot of land. Harry, Hermione, and Ginny are practically members of the family.

Our pigs are members of the family.

Common Pig Breeds

When you're choosing a breed of pig to raise on your farm, it's important to ask yourself a few questions. For example, is rapid growth important? Or is personality also a factor? Do you like your pigs extra fat? Or extra lean? Because we raise our pigs for meat, almost all our decisions are culinary based. I want to know that my pigs will give me lots of fat for lard and a big, thick, fatty bacon. I couldn't care less about the size of their loin, but I care very much about the size of their hams. If you're going to be raising pigs for breeding stock, disposition should most certainly be a factor. Here are some breeds you should consider:

▷ *Berkshire.* A common and high-quality meat breed of pig, the black Berkshire pig is distinguishable by its black coloring, light pink nose, and erect ears. This is a heritage breed that does well in outdoor setups. They'll top out at about 600 pounds at maturity.

Tamworth are wonderful backyard pigs!

▷ *Tamworth.* A very easy breed to spot, the Tamworth pig has a long and lean body that is covered in red hair. They have long noses and erect ears, and are great foragers. A full-grown adult will weigh about 550 pounds.

▷ *Hampshire.* Super easy to pick out from a group, the Hampshire is a black pig with a white strip that runs around its belly. Hampshire pigs are raised for their strong growth potential and excellent commercial carcass qualities–such as a thin layer of back fat (personally, I prefer a lot of back fat!). They average 650 pounds at harvest.

▷ *Duroc.* Like the Tamworth, the Duroc is a red pig. But unlike the Tamworth, Durocs have big ol' floppy ears that cover their eyes. Durocs are very hardy and have strong growth habits. The boars are commonly used to crossbreed. They average around 600 pounds at maturity.

▷ *Yorkshire.* The Yorkshire pig is the picture-perfect pig we tend to think of in America. Long, lean, and pink with erect ears, just like Babe!

▷ *Gloucestershire Old Spots.* I put them last on the list so you wouldn't think I'm biased. But I am. Old Spots are my favorites! Unlike the long and lean pigs that are so common, these pigs are large and fat and round. On top of their white hair and black spots, they've got large floppy ears and are so dang cute you could just hug and squeeze them! And I do. They are known for their foraging abilities, as well as their fat. Full grown, they'll weigh in at about 550 pounds.

Need-to-Know Basics

Again, genetics and diet will play a huge role in the harvest you can count on from a pig. That said, you can expect to harvest anywhere from 175 to 275 pounds of meat from a full-grown hog. That's a lot of pork, baby. We always scald and scrape our pigs, rather than skin them, which means we can harvest an entire hog with almost zero waste. Are you comprehending the awesomeness of pigs yet?

Housing

Pigs are fairly hardy and can do well in a variety of climates and weather. Naturally, you'll always want to offer your pigs protection from extreme weather and wetness. A safe, dry area during cold, windy, and wet weather will serve them well. Pigs actually build nests (yes, they're shaped like *gigantic* birds' nests) that they'll curl up in at nighttime. During hot months, the pigs will appreciate a cool mud pit to cool themselves in, or at the very least, a few spray-downs a day with the hose. Should I admit it's one of my favorite pastimes? Ahem.

Did you know that pigs build nests?

Feeding

On average, pigs will eat about 4 percent of their body weight each day. So if you've got a one-hundred-pound feeder pig growing in your backyard, you'll feed it four pounds of feed per day. Aren't you glad I'm here to help you with such difficult calculations? I'll admit something here: I never measure feed. Rather, I let my pigs enjoy as much food and garden scraps as I can possibly feed them. If I've got a lot on one particular day, I'll just scoop in a small amount of fermented grain. If I don't have many food

Pigs can make great use of garden waste and scraps.

scraps to share, I'll throw in a flake of alfalfa hay and a larger bucket of fermented grain. Pigs do very well on a large variety of feedstuff—garden scraps, gleanings from local orchards, waste produce from the grocery store, scraps from local bakeries or restaurants, or brewers' grains. We've never grown pigs faster than we did when we were feeding them extra milk from our dairy cow. Extra fat? Extra protein? Yes, please! They slurp up that milky goodness like a frat boy slammin' down a cold brew. And with about the same enthusiasm. If you're growing feeder pigs, you'll want them to have as much food as they want to eat!

Pasturing pigs is a great option as well and allows the pigs to dig for grubs, roots, weird small animals that creep around, bugs, and whatever vegetation they please. Some breeds are keener to graze than others, but all pigs can benefit from the space and room to, well, be a pig.

Premixed pig feed is available at any feed store and will contain trace minerals and optimal protein (around 18 percent) for growing a hog. Alternatively, a variety of grains can be mixed with food and garden waste as a maintenance diet for your breeding stock. It's important to remember that people have been raising hogs for thousands of years. Before we knew about amino acids and protein percentages, homesteaders were

FERMENTING AND SPROUTING GRAIN

As efficient as pigs are, they do even better when their feed is fermented or sprouted. To ferment the grain, cover it with water and allow it to sit for two days. To sprout it, allow it to soak up water for twelve to twenty-four hours before draining it, and then let it sit for a few days. Both processes make the grain more easily digestible and increase nutritional content.

throwing slops to their pigs. Moral of the story? Give your hogs a variety of high-quality foodstuff and they'll be just fine.

Watering

Every pig farmer may have his or her favorite watering system, but here on our farm, we like, ladies and gentlemen, the pig nipple. This metal "nipple" is attached to the side of a fifty-five-gallon, plastic drum and allows the pig to drink easily. It also alleviates the problem of them lying in and pooping in their watering trough. Which, trust me, they will. Different nipples at different levels of height will allow various sizes of pigs to drink easily.

Sample Annual Pig Calendar

Breed sows in December

Sows farrow in late March
(3 month, 3 weeks, 3 day gestation)

Wean piglets in late May

Raise for meat and harvest in December

Breed sows again in June

Sows farrow in September

Wean piglets in November

Raise for meat and harvest in August

Breeding Basics

Pigs are great candidates for artificial insemination if you can find a tech in your area who is willing to do it for you. Or you can have a fun weekend and go to a training yourself! Alternatively, you can keep a boar to breed your sows as necessary. There are many docile breeds available that can really be a wonderful addition to your small-scale operation. Our boar, Harry, is very well behaved and friendly. Do research in your own area to find out if there's a neighborhood boar that could breed your sows. Does your veterinarian perform AI on pigs? Do you have a facility to breed in?

Farrowing Requirements

When a sow is giving birth, she "farrows," while a cow "calves," a horse "foals," a rabbit "kindles," a goat "kids," and a sheep "lambs." Man, we farmers sure made this complicated, didn't we? Regardless, when a sow is farrowing there are a few things to keep in mind. For starters, the sow will need a safe pen away from the rest of the pigs. Little piglets are easy targets for other pigs to step on and chase about. A calm, warm shelter with thick bedding will be a much better environment for her to raise her little ones for a few weeks until they can grow in strength and agility! Left on their own in the wild, sows build large nests at the base of trees and under bushes. Keep the piglets dry, warm, and clean. Sows are notorious for crushing their piglets while lying down to feed them. Keep an eye on them for the first few days until everyone gets the hang of how to get out of big Mama's way.

Preparing for Butchering Day

A pig is a large animal. You're welcome for that incredible insight. My point is this: Harvesting an animal of this size takes a solid day of work for even the most experienced, so plan accordingly. Make sure that you've got your equipment at the ready. How will you hoist and move the pig? Do you have a place available to air-chill the carcass prior to butchering it? Just take your time and wrap your mind around how the process will work out, given the logistics of your space. We've done our pigs with the help of a hoist and also just with the help of my husband's incredibly strong biceps. Where there's a will, there's a way.

Harvesting and Butchering a Pig

Basic Equipment

.22 rifle

Sharp knife

Rope

55-gallon drum

Propane burner

Pig scrapers

Bone saw

Plastic bins and garbage bags, as needed

1. Use a .22 caliber rifle to shoot the pig in the golf-ball-sized circle right between the eyes. As always, the goal is to kill the pig as quickly and humanely as possible. As soon as the pig is shot, it will most likely convulse. At this point, stick the pig with an incredibly sharp knife in the artery that runs right behind the jowl. Stick the pig—don't cut. Those cheeks will be cured into pancetta, so don't mess with the meat. The stick will cut the artery and allow the blood to drain from the pig while the heart is still pumping. If you're super talented, you can use a bowl to collect the running blood for blood pudding. I'm not that talented.

2. Hose the pig down to remove mud, blood, and debris.

3. Hoist and lay the pig onto a large table or flat trailer. You'll want to be able to access the pig from all sides. Alternatively, you could use a pulley and gramble to hoist the pig from a tree branch or rafter in a barn.

4. Heat a 55-gallon drum of water with a propane burner until the water reaches around 200 degrees F.

5. Use small saucepans or buckets to dip into the hot water and pour over the pig, concentrating on one area at a time. Gently pour the hot water over the area and begin to scrape at the hair and skin with hog scrapers. If it's not coming off easily, keep pouring hot water over the area until the hair pulls out easily and the skin begins to scrape off quickly. Repeat this process with the entire pig. Once you get in the rhythm of it, you and a partner can probably scrape the pig completely in a matter of a few hours. When I was learning to butcher a pig, I asked my mentor if I had to get *all the hair* off. "That depends," he said. "What's your policy on hair in your food?" Use a sharp knife, if necessary, to scrape stubborn or remaining hair off the carcass. When you're done, if you've done it right, that pig will be smooth as a baby's bottom. I love the feel of a smooth pig! Don't you?

6. Remove the head. Gently cut through the flesh at the top of the neck until you hit the spine. Locate a vertebra and cut between it and the next one. Use your hands to twist the pig's head until it pops off, using your knife to separate any spare bits of flesh. Save the head to use for head cheese (see recipe on page 231).

7. Eviscerate the animal. The process is very similar for every animal. Be careful not to puncture the gut! A small slit on the underside of the belly up toward the tail will get you started. Continue cutting down the belly of the carcass with just the tip of your knife, making very shallow cuts. Using your hands, work the innards free. Everything that's in the body cavity needs to come out, by any means necessary. Have I mentioned that pigs are large? Be prepared to get in there, elbow deep, and get it all out, yo. Once the guts are free, cut a hole around the anus so that the rest of the intestines can fall out easily. Tying the anus off with a piece of string can prevent any leftover fecal matter from sneaking out during the evisceration process. Ain't nobody want leftover fecal matter sneaking into their pork.

8. Cut the pig in half using a bone saw, directly down the spinal column.

9. Rinse the pig of any debris.

10. Store the pig in a cool location overnight. This will make butchering the carcass into cuts much easier!

11. The following day, cut, wrap, and store the pork. Well done!

One pig can be enough to feed a family for a year! The work of raising them pays off big time.

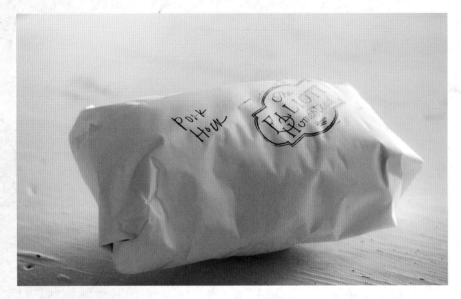

Waste Not, Want Not

As you continue to grow your own meat, you'll find yourself with a variety of extras that may leave you scratching your head. What to do with all these odd bits after the meat is harvested? Fear not. I've got some suggestions for you:

▷ *Heads.* Yes, I realize how gross this sounds, but animal heads are great simmered in a large stockpot with celery, carrots, peppercorns, parsley, and onions. Simmer on low for twelve hours, strain out the solids, and enjoy that rich stock in soups and stews.

▷ *Chicken feet.* Ditto to above. Chicken feet make delicious, gelatin-rich stock! I mean, you'll feel a bit like a witch doctor, but such is life.

▷ *Liver.* Hello, pâté. Welcome to the party!

▷ *Kidneys and hearts.* Delicious to eat fried in a skillet with butter, salt, and garlic. Not into it? At least feed them to your dogs, cats, or chickens!

▷ *Blood, feathers, hides, and intestines.* All can be buried in your compost pile and used to fertilize your garden next season!

How to Peel Chicken Feet

1. Place the chicken feet into a pot. Add enough filtered water to cover them (if a few toes are stickin' out, that's fine).

2. Put the pot on the stove and bring to a low simmer. Simmer the feet for 10 minutes.

3. After simmering, quickly move the pot over to your kitchen sink and run cold water onto the feet. Keep running the water for a few minutes, allowing the feet to "blanch" in a way.

4. Drain the water and move the pot to the counter. Using your fingers (the best tools ever created!) begin to peel away the skin. It's a bit slippery, but that's okay. I find that a twist and pull method seems to work best on the toes. Some people leave the toes on. Some clip them off. Do what you wish. Personally, I'm a toes-on kinda gal. Mostly because I'm lazy. I'm also not a perfectionist when it comes to this task. I don't mind a few bits here and there.

5. After peeling, the feet can be simmered for 12 hours with a tablespoon of vinegar, a chopped onion, a few stalks of celery, a few carrot sticks, and a lot of water to make a beautiful chicken-foot stock.

Scotch Spiked Chicken-Liver Pâté

8 tablespoons high-quality butter

1 onion, minced

3 cloves garlic, minced

6 tablespoons fresh herbs of choice

1½ pounds fresh, homegrown chicken livers

1 tablespoon scotch whiskey

Sea salt and pepper, to taste

½ teaspoon lemon juice

1. Melt the butter in a skillet.

2. Add the onion and garlic. Sauté until fragrant and golden, about 10 minutes. Might as well toss in those herbs now, too.

3. Add the liver. That's it now, don't be afraid. Sauté for 5 minutes per side or until just cooked through.

4. Drizzle in that delicious whiskey. If you don't have whiskey, a tablespoon of white wine would also work nicely.

5. Let the pâté simmer for a few minutes while all those delicious flavors mingle. Then, transfer to a high-powered blender or food processor. Puree until smooth, scraping down the sides if necessary.

6. Season to taste with salt, pepper, and lemon juice.

7. Chill the pâté until you're ready to eat. I like it best chilled, but I'm sure some like it best warm. Rumor has it that pâté tends to taste its best after a few days' time in the fridge (ours never lasts that long).

Storage and Preservation of Meat

Imagine living back in the day when refrigeration wasn't an option. Here you'd stand, with two hundred pounds of pork, and no cool place to store it. What would you do? I'll tell you what you'd do. You'd get to curing. Salt and smoke are the two primary forms of curing still familiar to us today. The man who taught me how to butcher a pig, an incredibly talented local meatsmith, taught me early on that *salt is magic.* Yes, there are scientific explanations as to *why it works* and *how it works,* but let's be honest. It's just magic. Water is what causes meat to spoil and salt pulls water from the meat, protecting meat from spoilage. Beyond that, salt turns meat into something completely different and complex. Salt-cured meats are the best. There. I said it. Once I'm really cool, I'll be able to cure an entire pig without utilizing the freezer at all. Until then, I'll cure what I can and freeze the rest. As with all cuts of meat, they should be thoroughly wrapped in plastic wrap and butcher paper before being labeled and stored in the freezer.

Pork roasts can easily be cured into hams right in your own kitchen!

Homemade Ham

Don't let the thought of curing your own ham at home deter you from enjoying such awesomeness. As with all things home-made, a little bit of extra effort pays off ten-fold in taste and experience. Though it takes a bit of time to cure, the active work time to home-cure a ham is roughly 0.192334 seconds. You can do it! I have the utmost faith in you.

1 fresh pork roast, preferably from the leg

Sea salt

Whole dehydrated cane sugar

Kitchen scale

Large bowl

1. Dry the ham with a rag to absorb any extra moisture. Weigh the roast and write down this number.

2. For every pound of meat, weigh out 15 grams of sea salt and 2.25 grams of sugar.

3. Rub the meat with the salt/sugar mixture. Get into every crevice that you can find. Really work it in there. If you have a bone-in roast, pay special attention to that area. Make sure that the roast is coated.

4. Place the roast in a large bowl, uncovered, and set it in the refrigerator. Measure the diameter of the roast and add 3 to that to figure out how many days it will cure. If your roast is 7 inches in diameter, then 7 + 3 = 10 days total for curing. Let the roast be. The salt will do its magic.

5. After the proper curing time, remove the ham. At this point, the ham can be baked and glazed with whatever you'd like. Honey and molasses are my favorite! If my husband has his say, the hams are always smoked.

6. Smoke or cook the ham to an internal temperature of roughly 170 to 180 degrees F.

Homemade Bacon

Back in the day, people knew how to do cool things. Well, I think they're cool. Surely I can't be the only one in the world that thinks it's exceptionally cool to cure pork belly and hang it in your kitchen.

Fresh pork belly from the best-quality hog you can find (if you can't grow your own pigs yet, find a local pork farmer or talk to a high-quality local butcher)

About 6 cups dehydrated whole cane sugar

About 6 cups coarse sea salt

1. Combine the sugar and salt. Generously rub the flesh side of the pork belly with the mixture.

2. Rub the sugar and salt into the flesh some more.

3. Did I mention you need to rub the ol' pork belly down with the sugar and salt? Make sure to get the sides too—anywhere water can accumulate.

4. Stack the uncovered pork belly into a large plastic bin. Stick it in the refrigerator and forget about it until the next day. Dump the accumulated liquid out of the plastic bin and rerub the flesh with the sugar and salt. Stack it all back in the tub and stick it in the fridge again.

If anything is more delicious than homemade bacon, I've yet to taste it.

5. The next day, dump out any accumulated liquid and rub the sugar and salt mixture on any part of the pork belly where the salt and sugar has completely dissolved. A thin layer will do. Repeat this process every day until liquid stops accumulating in the bin. On average, around a week or two will do.

6. Rinse the pork belly with water, using your fingertips to scrub off any remaining sugar and salt. Pat dry.

7. Voila! Cured bacon.

8. At this point, you can run a meat hook through a corner of the bacon slab and store it at room temperature while you cut off pieces to cook up as you wish! We bring out our slab on special mornings, cut off a piece, and fry it up. Everyone loves bacon day.

Homemade Chorizo

I love it when the chorizo gets crispy and brown in the frying pan. That's my kryptonite right there.

3 pounds ground pork, at least 30 percent fat (aka: sausage meat)

1 large onion

3 tablespoons balsamic vinegar

1 tablespoon cumin

1 tablespoon coriander

2 bay leaves

5 cloves

2 allspice berries

¼ teaspoon cinnamon

1 tablespoon fresh oregano

1 tablespoon fresh parsley

1 teaspoon salt

1 dried chile (or ½ teaspoon red pepper flakes)

½ teaspoon black pepper

1 tablespoon sweet paprika

1 tablespoon smoked paprika

5 cloves garlic

Red wine

1. Put the ground pork in a large bowl. Set aside. Ignore the pork. Let the pork be.

2. Combine the onion, vinegar, cumin, coriander, bay leaves, cloves, allspice berries, cinnamon, oregano, parsley, salt, chile, pepper, paprikas, and garlic together in a food processor or high-powered blender.

3. Blend the ingredients until smooth, adding red wine as necessary to ensure it all gets blended well together.

4. Combine the sauce and pork. Mix together.

5. Oh, by the way, that's it! You've now made chorizo. All that's left is to fry it up and enjoy! You can cook it as is, mold it into patties, or put it through a sausage stuffer to shape it into links.

Homemade chorizo is delicious mixed up with garden potatoes and radishes!

Rillette

Fat's where it's at, baby. Rillette takes all the little trimmings left over from butchering and cooks them to perfection in their own fat. Think of it like a country pâté!

**Pork scraps from butchering
(a combination of meat and fat of any ratio will do)**

Sea salt and freshly ground black pepper, to taste

Dutch oven

1. Cut the pork scraps into 1- to 2-inch pieces. Add them to the Dutch oven and turn the heat to low. Low, I say. Cover with the lid.

2. Let the pork slowly cook and melt, stirring as often as you remember to. We're not looking to sear or brown the meat. Just gently, slowly, patiently, let the pork melt and cook on its own.

3. Once the pork bits are nicely browned (I once did about 3 pounds of scraps and it took about 20 hours), turn the heat off, and gently remove the bits of meat from the Dutch oven, using a strainer to skim through the fat. Place the meat bits into a large bowl and season to taste with salt and pepper. Give it a little extra salt, as you'll be eating it at room temperature.

4. Use two forks to gently shred the bits of meat, breaking up any large chunks.

5. Using a set of tongs, scoop the meat bits into a large mason jar, pushing down with the tongs to help release any air bubbles and condense the meat. Once all the meat is added, pour the melted fat left in the Dutch oven over the top of the meat, covering it by at least ½ inch. The fat will seal the meat and preserve it. Place a two-piece lid onto the mason jar and immediately stick the jar into the refrigerator. This will cause the jar to seal as it cools (essentially "hot-packing" the meat). Once the jar has sealed, you can store the rillette in your refrigerator for many months. Smear it on a piece of bread and enjoy.

Rillette is easy to make and will last for months in your refrigerator.

Head Cheese

Now, don't go gettin' all grossed out on me. Just because the meat didn't come from a super familiar part of the body, like the pork belly, doesn't mean that it's gross. It just means that you're not used to it. And as a foodie, might I just point out, it's always worth trying something before you turn up your nose at it.

1 pig's head, cleaned

Water

Sea salt and pepper

Large stockpot

1. Put the head (frozen or fresh) into a gigantic stockpot (like a seriously large pot). Cover with filtered water.

2. Slowly bring the pot to a very low simmer. Cover and allow the pig head to simmer on low for 24 hours. By the time the head is done, it will be falling apart into pieces.

3. Carefully remove the head from the stockpot and place it onto a large platter (reserve the cooking liquid). Let it cool before using your hands to pick the meat from the bones.

4. Place the meat into a large bowl. Season to taste with sea salt and pepper. Err on the side of a little salty, since head cheese is typically eaten at room temperature or cool.

5. Bring the cooking liquid back up to a simmer. Continue to simmer the stock until the liquid has mostly reduced and is slightly thick.

6. Place the shredded meat into a pan (like a loaf pan). Pour the reduced liquid over the meat until the pan is full and the meat is submerged.

Head cheese makes use of a delicious cut of pork that's often overlooked.

7. Refrigerate the head cheese until it is set. The liquid will set and result in a gelatinous loaf of the tenderest meat imaginable. Serve the head cheese sliced cold with a nice, crusty loaf of bread.

As much as I love our pigs, I'm also thankful to have a few friends on the farm that require less work and smell much better! After all, you know what's heavenly drizzled over homemade ham? Your very own honey!

CHAPTER 6

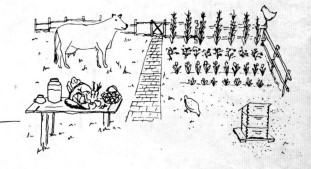

Beginning Your Apiary

Why keep bees? Bees, in my humble opinion, are some of the most important insects on the planet. We need them for *our survival*. How's that for importance? Without bees, crop pollination simply wouldn't happen. A honeybee hive should be a welcome addition to any backyard homestead, as the bees will not only pollinate the crops in your garden, but also give you an abundance of raw honey to drizzle over your strawberries, stir into your morning coffee, and whip into your egg-white meringues. I find that keeping honeybees is the most intriguing endeavor on our little farm. They continually teach and inspire me! This may sound silly. After all, they're "just" bees, but after spending time with my hives for the past few years, I've come to appreciate the bees' complexity and social structure. Did you know that a worker bee will actually work itself to death? By the end, its wings will be tattered from hundreds of miles of travel over its forty-day life. I'm feeling some similarities to motherhood here . . .

Getting Started

To get started you'll need some bees, of course. And yes, you *buy* a package of bees in the spring from a local supplier. (Some hives are started with captured swarms, but that's a whole other story!) Each shoebox-sized screened package weighs about three pounds and will include a queen and roughly ten thousand bees!

The Bee Cast of Characters

Adult: On day twenty-one, the adult bee emerges from the cell of the comb and begins its work!

Drone: Male that mates with queen

Egg: Remains safe in the comb for three days before turning into a larva.

Larva: Fed by nurse bees for the next six days until they reach full size.

Pupa: The larvae transform to pupae, which takes ten days.

Queen: Egg-producing female

Worker: Non-egg-producing female

Bees can be purchased through a local beekeeping chapter or online retailer.

Alternatively, you could start with a "nuc" (pronounced "nuke"). A nuc is four to five frames of comb and bees from a working hive, including the queen. Some people prefer to start with nucs because the bees get off to a faster start, since they already have some frames drawn out in comb. Naturally, nucs cost more.

There are essentially two types to choose from:

1. *Italian bees.* These lightly colored bees are highly adaptive, less prone to diseases than other European breeds, extremely productive, and known for their extended periods of brood rearing.

2. *Carniolan bees.* Probably the most popular honeybee around, Carniolan bees are easy to work with, strong wax producers, and reproduce rapidly (leaving them more susceptible to overcrowding and swarming).

Equipment

For anyone looking to start with just one hive, here are the basic pieces of equipment you'll need to get your bees a-buzzin'!

The hive. Mostly commonly composed of a bottom board, hive body, honey super, inner cover, and outer cover, the hive is essential for allowing your bees a safe place to breed and store honey. The Langstroth hive is the classic rectangular-shaped hive that you'll see nestled among orchards and gardens. It's become the benchmark for the standardization of beekeeping.

Another common type of hive is the top-bar hive. This style of hive allows the bees to naturally draw the comb to the size and shape they wish, as they would in nature, while gently persuading them to draw the comb down from the top bars laid across the top of the hive. Because the comb is not as standardized as it is in the Langstroth hives, harvesting can often mean cutting the comb off completely in order to squeeze the honey out, whereas in the Langstroth hive, the frames of honey are spun in a honey extractor to release the honey while keeping the wax intact. Each beekeeper builds and structures their hives a bit differently. That's what makes it so unique and fun!

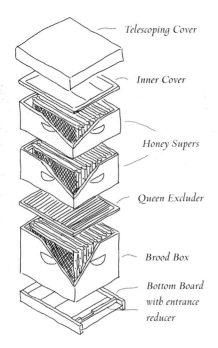

Telescoping Cover

Inner Cover

Honey Supers

Queen Excluder

Brood Box

Bottom Board with entrance reducer

Langstroth hive

Hive Cover

Top Bar with drawn out comb

Hive Box

Hive Entrance

Top-bar hive

The frames and foundation. While the top-bar hives allow the bees to build their foundation from scratch, the Langstroth hive is assembled with ten frames per super, each of which has a foundation that the bees will build their comb on. The idea of the foundation is to encourage the bees to build a particular shape of comb so that it's easy to harvest. The comb will be drawn out from the foundation by the bees and will house larvae and store honey.

Bees will use the foundation to draw out comb and fill it with honey.

The suit. Obviously, if you're going to work with bees, having a suit is super helpful. Ain't nobody wants to get stung by bees. I'm semi-sorta-kinda afraid to get stung by our honeybees, so I'll always suit up. Plenty of beekeepers are comfortable working their hives without one, but call me a sissy and gimme my suit.

The smoker. When a keeper opens a hive, guard bees emit pheromones to let the hive know that it's under attack! Smoke will mask this pheromone and allow you to work your hive with less stress to the bees and most certainly fewer stings for you!

I was once told, "If you don't have time to properly light your smoker, you don't have time to properly check your bees." Turns out, that's very true.

The queen excluder creates a barrier that the queen cannot fit through, keeping larvae from ending up in your honey harvest.

The hive tool. A small metal tool that is invaluable to have in your suit pocket to open up the sticky hives and to separate the frames for inspection.

The queen excluder. This is a thin, metal sheet that fits right over the top of your brood box. It is filled with holes big enough for most of the bees to fit through, but not the big ol' queen. This means that she's confined to the bottom box and unable to crawl up into the top honey super. Result: No larvae in your honey harvest.

The bee brush. Traditionally, goose feathers were used to gently wipe bees off the frames for inspection and harvest. How delicate! Here on our farm, we use a natural-fiber bee brush, which is specifically designed to avoid crushing the bees' legs as they're wiped off.

How to Hive Bees

If you're just getting started, you'll most likely be receiving a package of bees in the mail. The first step in starting your apiary will be to get that box of bees into your hive. Plan ahead a wee bit so that you can have your hive equipment ready and (surprise!) a mini marshmallow on hand (you'll find out why in a minute).

1. Acquire a package of bees. Each of these packages will include one queen that will come in a separate small box within the package. This is to keep her safe from the other bees while they adapt to her phero-mones and begin to accept her as their queen.

2. Prepare your hive. Is your hive all ready for the bees? It should be before they arrive. Have it set up in a safe location, lifted off the ground a bit, where it will get protection from severe weather or too much sunlight. We keep ours at the base of an old oak tree so that the branches can provide some shade and shelter from weather. Once you hive your bees, you won't want to move them, so put a bit of fore-thought into where to set your hive. To begin, you'll only need one brood box filled with ten frames. Too much space and the bees will get confused! Rather, let them build their hive up one box at a time.

3. Suit up!

4. Spray the package of bees down with a one-to-one ratio of sugar and water in a squirt bottle (see recipe on page 244). A few spritzes will do the trick.

5. Slam the box of bees down on the ground three to four times to knock them all to one side of the box (don't worry, they cushion each other's fall).

6. Remove the queen, still in her cage, and set her aside. A pocket is a great place to keep her for the time being.

7. Remove the lid from your hive. Remove three frames, so there are just seven remaining. If you don't remove the frames before dumping the

bees in, they'll have nowhere to go and will instead just pile on top of the frames, so don't skip this step!

8. Dump and shake the bees into the hive through the open hole in the package. They'll all fall to the bottom in a giant pile. Don't worry. That's what they're supposed to do. Continue to shake the bee box until you get as many bees out as possible.

9. Gently put the three removed frames back into the box. The bees will slowly move out of your way. So now you should have one box with its ten frames back in. With the bees inside.

10. Now, let's turn our attention back to the queen that's in your pocket. You'll notice that her little cage most likely has a small hole that's been plugged with a cork. Using a screwdriver or knife, gently pry the cork out. Quickly put your finger over the hole so she can't climb out. Then, carefully, stuff the hole in the queen's cage with one mini-marshmallow. I know, this might seem really weird, but it's the perfect way to introduce the queen to the hive. By the time she eats through that marshmallow (two to three days), the others bees will have accepted her presence, and she will release herself from the cage. Some people prefer to manually release the queen two to three days after hiving the bees, but I like the marshmallow method.

11. Now that the queen has her marshmallow, gently wedge her cage in between two frames. Make sure the hole is facing in a direction that she can climb out of once she's eaten the marshmallow.

12. So here we are—one hive box, ten frames with foundation, three pounds of bees dumped inside, one queen in a cage that's plugged with a marshmallow. Are we good? Everyone with me still?

13. The last step in hiving the bees is to give them some sugar syrup to drink while they build their comb. Because they don't have comb yet, they don't have a way to store any food. Because of this, it's important to supplement them until they have time to build up their hive, store food, and raise new bees.

Get plugged in with a beekeeping organization in your area. They'll give you expert advice on how to set up your hives and care for your bees.

Homemade Sugar Syrup

One part sugar

One part water

In a small saucepan, combine equal measurements of sugar and water until dissolved. Pour the mixture into a mason jar. Use a hammer and nail to poke a few small holes (think the tip of a pen) into the top of the metal lid and screw it on the top of the mason jar. Tip the jar upside down and place it on the telescope cover of your hive. The bees will be able to access the syrup from below in the hive. I like to put an empty super around the jar to keep the wind from tipping it over. Ideally, this will be the only time you'll need to give your bees this concoction.

Checking Bees

Checking bees is a routine part of beekeeping. During the summer, which is the time of year when a hive is at its most productive, it's advisable to check your hive at least once a week to watch for signs of swarming and to prevent it from happening. When bees swarm, they take half of the population with them, leaving you with a lot fewer bees (read: a lot less honey!). Routine checks will also help you identify the strength of your queen, watch for any signs of disease, and take a peek at your honey supply. Life in the hive during the fall and spring is typically not as active, so you can space your inspections a bit more if you desire (though many beekeepers will tell you the temptation is always to open the hive more often). During the winter, let the bees remain closed up in the hive to keep them from getting chilled as their resources and energy are limited during that time of year. Only open the hive if the temperature is *at least* in the sixties and it's a warm, sunny day. You can always give the side of the hive a few knocks with your knuckles to make sure things are still buzzing on the inside.

Checking the Hive Checklist

After suiting up, lighting your smoker, giving the hive a few puffs of smoke, and opening the top lid of the hive, you should check for the following:

▷ *Signs of life.* Yes, there should be bees moving and flying all around. If there are no bees, that ain't a good sign. Take a few minutes to watch your bees. Are they active? Moving normally?

▷ *Honey stores.* The top box of your hive will be your honey super, that is, where the bees are storing their honey. How's it looking? Do you see lots of nicely capped honey? This means they're doing their job.

▷ *Enough room.* Bees need room to store honey. Without enough room, they'll swarm and half of the hive will leave in search of a new home. Bees typically will fill the middle of the box with honey and brood before moving their way toward the edges. Your bees should always have at least a couple of empty frames that they can work on filling. Any less and they'll start to prepare for takeoff! If room is needed, now's the time to add another box with fresh frames.

▷ *Signs of swarming.* The biggest indicator that your hive is about to swarm is the sign of queen cells built along the top edge of a frame. These queen cells are built bigger to house the larger queen and look like peanut shells. When these cells are lined up at the top of a frame, it indicates your hive is preparing a new queen for swarming. Pinch these cells off and make sure your hive has enough room to prevent them from doing it again.

HIVE TAKEOVER

Queen cells are sometimes built on the bottom edge of a frame. These queen cells are typically built when a hive is creating a new queen (a "supersedure") to take over and replace an old or ailing queen.

▷ *Brood.* Any time you open your hive, you should see various stages of bee larvae in the brood box. Take note of the pattern in which the queen is laying and where the brood is located in the box, which are all indications of queen health. Ideally, a queen will lay in each cell, creating solid patches of larvae. If she is randomly laying around the frame and is leaving lots of blank cells, it may be time to replace her with a more vigorous queen!

▷ *Disease.* See any intruders? Mites? Ants? Any funky molds? Take note of anything out of the ordinary.

▷ *Queen spotting.* Easier said than done, it's always a treat to get eyes on the queen while checking your hive. It's is reassuring to know that she's in there and doing well. The queen has a much longer and narrower abdomen than the other bees, making her stand out. Typically, other bees will create a circle around her with their faces toward her, making her easier to spot.

Bees should be checked routinely for activity and health.

The honey can easily be capped with a hot knife.

Harvesting Honey

The first time we ever harvested honey from our bees, I was elated. Actually, even the word *elated* doesn't do justice to the incredible feeling I experienced. With so many hard lessons learned in homesteading, to have a new venture that actually went, dare I say, *as planned*, was a huge reward. We harvested gallons and gallons of honey that first year and as we bottled it into glass jars, our entire family was beaming. We spent the rest of the night licking the sticky stuff off our fingers and spoons, and no one complained. This is by no means the only way to harvest honey, but it's how we roll around here. Here's how you do it:

1. *Get the equipment*. A centrifuge honey extractor is what you'll need. It's a fairly simple device. You insert the frames of honey inside the mechanism, and with a rapid turn of the handle, the extractor spins

A honey extractor spins the hive frames to extract the honey via centrifugal force.

them incredibly fast. The honey spins from its comb and drips down the side of the extractor, to be bottled through a valve at the bottom. Pretty straightforward. What I like about this method is that it leaves the comb intact. If you choose to harvest the comb for wax (which would no doubt serve a great purpose in making homemade beeswax candles), it would require the bees to rebuild that comb before they could once again fill it with honey. On our farm, we raise bees for the honey, so I always try to leave as much comb for them as possible.

THE SHELF LIFE OF HONEY

Did you know that honey never goes bad? Ever! It can sit on your shelf for years and years and remain perfect. Reason #198,273 to love your bees.

2. *Cap the honeycomb.* Once you're ready to harvest the honey, use a warm knife to carefully and gently uncap the honey. This is done by sliding the blade of the knife across the comb, scraping off the layer of wax that caps the honey. Once the honey is uncapped on both sides of the frame, it can go into the extractor.

3. *Spin, baby, spin.* Spin the handle on the extractor continuously. Work it! You can stop, pull out the frame, check your progress on how much honey is still left in the comb, and then keep on going.

4. *Lather, rinse, repeat.* Once the frame is completely empty on one side, turn the frame and repeat the process on the other size.

5. *Filter and bottle the honey.* After extracting the honey, simply run the honey through a small, mesh strainer (it helps to catch any chunks of wax or dead bees that have fallen into the honey) and then pour it right into a bottle or jug for storage.

Honey Meringues

Natural sweeteners, such as honey, have a lot more flavor than processed white sugar. They lend something besides just sweetness—an essence, a presence, a new level of taste. And when you're baking simple sweets, that's how magic is made. These honey meringues are out of this world and almost always in our kitchen.

4 egg whites

1 cup honey

1 teaspoon vanilla

Generous pinch sea salt

1. Start by removing the bowl from your stand mixer and placing it in the refrigerator until it's nice and cold (about 20 minutes). Egg whites whip up quickly when the bowl is cold!

2. In another bowl, combine the egg whites, honey, vanilla, and sea salt.

3. Set the bowl over a medium-sized saucepan that has been filled with about 3 inches of water. Bring the water to a low boil. While the water is boiling, begin to whisk the egg whites and honey together with a large whisk. Continue to whisk for 3 minutes until the egg whites have aerated and turned into a nice, white froth.

4. Pull the chilled bowl from the refrigerator and attach it to your stand mixer with the whisk attachment. Pour the frothy egg mixture into the cold bowl and begin whisking on high. Continue to mix until the meringue is shiny and has stiff peaks.

5. Line a baking sheet with parchment paper. Scoop spoonfuls of the meringue mixture onto the sheet. You can do 'em fancy . . . or sloppy . . . or somewhere in between. I won't judge you on your meringue skillz, I promise. I've always been more of a fan of "rustic" desserts anyway.

6. Place the meringues in a 225-degree oven. Bake for 1½ hours. After this time, shut the oven off and allow the meringues to sit in the warm oven to dehydrate a bit until they're the perfect crispy, chewy combination of your choosing. I usually let mine sit overnight and enjoy them the next morning . . . or, let's be honest here, decide to just eat them gooey right off the baking sheet without wasting any time at all (cough). Self-control? What's that?

7. For even more delicious meringues, top them with a bit of melted chocolate or homemade preserves swirled in before baking. Heaven.

Homemade Beeswax Candles

Considering that synthetic wax candles can be purchased at the store for next to nothing, this certainly isn't a project that's done for cost savings. That's not the point. The point is that there's something wholesome about it . . . something grounding . . . something more valuable than just the candle itself. Bonus: Did you know that beeswax produces a negative charge when it's burned? And did you know that pollen, dirt, and dust are charged by positive ions? Thus, burning a beeswax candle actually causes these particles to drop out of the air, thus purifying the air? Now you do. Onward to candle making!

You will need:

> 3 to 5 pounds 100 percent pure beeswax
>
> Deep container, such as a metal tin or glass jar
>
> Stockpot
>
> Bowl
>
> Water
>
> Wick
>
> Wax paper

You can turn your extra beeswax into room-purifying candles.

1. Roughly chop the beeswax into large pieces. Place into your deep container (I use a glass gallon jar).

2. Over medium heat, heat a large stockpot full of water.

3. Place the deep container into the large stockpot. This creates a double boiler of sorts and allows the beeswax to melt gently in the heat of the water.

4. Cut a piece of wick to double your desired length. For example, I usually do 8-inch taper candles, so I cut a 16-inch piece of wick. You will dip both ends of the wick into the beeswax and you can easily hold it up by the middle.

5. Once the beeswax is melted, pinch the wick in the middle and let the ends of the wick hang down. Slowly dip the ends into the wax. Then, dip into a bowl of water (this will set the wax). Shake off any excess water and dip once again into the wax. Repeat the water-wax dipping cycle until the candle reaches the desired thickness.

6. Cut off the bottom inch of the candle to create a straight bottom. Dip once more for good measure.

7. Hang or lay on wax paper to dry. Let set for 24 hours before burning.

There's just something romantically beautiful about a collection of fresh, homemade beeswax candles lining the kitchen counter. And the smell? Fuggedaboutit. It's a little piece of air-freshening heaven on earth.

CHAPTER 7

Homestead Orchard

I must admit, I'm pretty spoiled in the fruit department. Not because I have a super-successful homestead orchard myself, but because I live in orchard territory. Our farm is surrounded by cherry, apple, and pear trees. Peach, nectarine, apricot, and plum orchards are easy to source from during the harvesting season, and our kitchen counters become loaded with boxes and bins of all sorts of incredible fruit. As I type this, no less than one hundred pounds of boxed peaches are littering my kitchen floor awaiting their destiny to mingle with honey syrup in mason jars. These jars will be lined up in our root cellar, along with a variety of other canned goods, for winter sustenance. Orchards, to me, conjure up memories of my childhood spent with my grandpa harvesting pears from the orchard that filled up the space behind his home. They remind me of summers spent as a teenager lugging crates of cherries through the rows of trees, laboring in the blazing summer sun (hello, manual labor!). They remind me of a Frenchman telling Anthony Bourdain to put his nose in the box of peaches to smell and inhale their magnificence.

Even if you live in suburbia, having a few fruit trees planted on your property is the most wonderful way to enjoy the succulent tastes of the season. It's amazing the amount of fruit that one tree can produce, filling your family's bellies each season and beyond. Can't you picture it? Baskets and baskets full of pears. Pear cobbler. Dehydrated pears. Canned pears. Pears poached in red wine. May your imagination know no limits.

I'll admit that I don't have nearly enough fruit trees on our small farm yet. Currently, only a pear tree and a plum tree reside by my whiskey-barrel planters (filled with lavender, naturally) at the entrance to our

Cherry and pear orchards surround the farm on every side.

potager. They're new–barely more than a year old. But I dream of the day that their branches sprawl over the pea-gravel pathways and shade the lettuce planted in alternating colors.

Even though I grew up in orchard country, it wasn't until I watched the BBC's historical series *Edwardian Farm* that I really fell in love with the idea of foraging, growing, and preserving fruits. Historian Ruth Goodman would take her wicker basket out into the hedgerows of England on the hunt for crabapples, currants, or plums. She'd toddle back to the vintage stone farm, baskets in tow, and mix up a chutney or jam for the larder. That intentional action toward preserving and enjoying a fruit during its particular season speaks to my soul. And so, we grow. We plant a tree in the hopes that we will gather from its branches, eat from its offerings, and that long after our days have passed, others will taste of its fruit and feel it all as well.

Before You Plant

Here are some tips before you start buying trees and digging holes:

1. *Source the right tree for your area.* This is a big one. Many trees have "chilling hour" requirements–meaning the amount of hours under 45 degrees each year they require to produce fruit. If you plant a tree that requires 150 chilling hours and your zone only gets 75, well, that would surely stink, wouldn't it? Often local nurseries will only source rootstock that has been preselected for your given area, taking out the guesswork. However, if you're purchasing from an online retailer or an out-of-town nursery, take note of zone, chill, and soil requirements.

Keeping the tags attached can help you remember which variety you're growing and its specific requirements.

2. *Map out your available space.* Trees get big! And unlike your tomato plants, you won't be ripping these up each fall to "try again" next

FRUIT TREE SIZES

Dwarf: As the name implies, this is a variety of fruit tree that is bred to take up less space than a standard fruit tree. Typically, dwarf trees average around six feet tall at maturity.

Semi-dwarf: Slightly larger trees than a dwarf variety, these trees average around twelve feet tall at maturity.

Standard: Large trees, averaging around twenty feet tall at maturity. Imagine the ladder you'll need for that bad boy!

year. Nope. If you're planting fruit trees, you're in it for the long haul. There are dwarf and semi-dwarf varieties commonly available, which will reduce how much space you need to allot for each tree. Make sure that once the tree's full grown, it won't shade your garden beds (unless you want that) or run into a building. Trees need room to breathe, man.

3. *Pay attention to pollination requirements.* Each variety and species of tree has different pollination requirements. Some are self-fertile, meaning

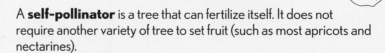

CROSS-POLLINATION VS. SELF-POLLINATION

A **self-pollinator** is a tree that can fertilize itself. It does not require another variety of tree to set fruit (such as most apricots and nectarines).

A **cross-pollinator** is a tree that cannot fertilize itself. It requires pollination from another variety in order to set fruit (such as apples and sweet cherries).

they don't require another tree for pollination, whereas others require another tree of the same (or different) species for pollination. Make sure you pay attention to this. There's no point planting a beautiful pear tree if you don't have another tree around that can pollinate that awesomeness!

4. *Consult your county extension office.* They'll give you advice about regulations on growing fruit trees and maintenance requirements of fruit trees in your area.

5. *Dream big!* Life's too short to be afraid of the years of investment a tree needs before it will bear you fruit. You can wait three years, right? Just get on it already! There's no time like the present. And the world will be a better place for it.

Bees are essential for pollinating all those beautiful little blossoms! If you're planning your homestead orchard, by all means, add some bees into the mix! Pollination for your trees and honey for you! Heck yes.

Planting Fruit Trees

Before you plant your fruit trees, I'm going to advise that you take a step back. Monitor where the sun hits during the long and short days of the year. Watch where the trees should be planted to avoid the shade of other trees and where a tree that needs full sun should make its home. Watch where the water drains and where the soil is susceptible to dry out. This is all extremely valuable information for the home orchard and will prevent you from wasting years of trying to grow trees in the wrong spot.

Let's be honest, I'm most certainly not the most patient person in the world, but with planting the home orchard, even I can show a teeny bit of restraint. It's worth it. Because the only thing sadder than waiting a bit while monitoring the property is watching your beloved tree shrivel up and die. Most fruit trees require lots of sun, so make sure to observe both winter and summer sun situations on your property before choosing a permanent spot for your tree.

To Plant

1. Dig a hole that's three feet across and at least a foot deep. This helps to loosen the soil and allow the roots to spread easily as they seek to establish themselves. Feel free to con a strong friend into doing said work for you, if need be. Shoveling is hard, man.

2. Prepare the fruit trees for planting by removing them from their containers and laying them out. With bare-root trees, the roots are typically wrapped in some wet wood chips and newspaper to keep them moist during shipping. If your trees come in plastic nursery containers, break up the root balls slightly before planting.

3. Look for the root line on the trunk of the tree. This is the line where you will visibly see the color change. You don't want to bury the tree any deeper than this line. Spread the roots out in the hole to encourage growth in a variety of directions. Fill the dirt back into the hole and gently press around the tree trunk. It's important to fill the hole

back up with the native soil for a few reasons. For one, soil that's too rich will encourage the tree roots to stay right where they are–in that little baby shallow hole–instead of venturing out into the wide, underground world like we want them to. And two, it helps to build a strong and healthy tree that is capable of withstanding hardship. Rich soil equals a sissy tree. Think of it as tough love. A bit of lime can amend acidic soil and bone meal is a welcome addition for healthy root development.

4. If your fruit trees came from nursery containers, go ahead and give the transplants a bit of water to get them settled nicely. If they were bare-root trees, watering is usually not necessary for at least a few days until the tree wakes up a bit from its dormancy. Alternatively you can soak the bare roots in a bucket of water for a few days to perk them up before planting.

5. Mulch around the tree heavily to help with weed suppression and water retention. Once the tree begins to grow leaves, you want to make sure the soil stays slightly moist (but not soggy!) at all times. Trees have an incredible ability to spread their roots and find water, so take it easy. More diseases and death in fruit trees come from overwatering than underwatering.

Mulch can help keep your trees protected as they grow.

6. Wait a decade for your fruit trees to begin bearing fruit. Just kidding! It's not that long. Though, I'm sure it will seem like it. Growing fruit trees on the homestead is not for the impatient. It's usually at least three years before you can expect to harvest fruit. That said, there's hardly a better investment you could make in your homesteading future. Once established, fruit trees will bear fruit for decades. For your generation, and the next, and maybe even the next. How fantastic is that?

Care

Backyard fruit trees are fairly easy to maintain. They thrive in balanced soil and with deep watering, consistent pruning, and bit of protection from some common pests. You'll want to pay close attention to any signs of stress, disease, or pests in your trees. Orchardists in our neighborhood earn their living by the quality of their fruit. It's my duty, as the farmgirl, to grow mine to the best of my ability as well! If you get to the point of not being able to care for your tree, don't just let it become diseased, breed pests, and infest all the other trees in your area. Just pull it out. Trust me. Orchardists will thank you.

Set aside time to prune your fruit trees in the fall. They'll thank you!

Pruning

I have a confession to make. I'm not the best pruner in the world. I've had a few orchardists show me their tricks but every time I go to implement what I've learned, I end up doing what looks pretty instead of what will encourage strong growth. Does that mean I've failed as a farmer? Surely not. It means that I need to keep practicing and get over my desire for constant aesthetic appeal over functionality. I do know this much:

▷ Trees should be pruned in the fall.

▷ Old, dead branches should be removed, as should any branches that show signs of disease.

▷ Water sprouts (the fleshy, non-woody stemmed branches that shoot up from the big limbs) should also be removed.

▷ Branches should be thinned so that each branch has roughly 8 inches of air space around it for growth. Pinpoint the branches that look healthiest and are growing in the correct direction. Remove everything else. The goal is to have the branches spread out in an even and strong pattern. Or, if you're taking after me, they "look pretty."

▷ Branches should be cut back a bit to encourage strong, shorter limbs versus long branches that will easily break under the weight of the fruit.

Thinning

Pears and apples really benefit from being thinned. True story: When I worked as a florist, we would go out with our bags into the orchards every spring and gather as many baby apples and pears as we could possibly manage. Because they looked really pretty in flower arrangements. Thinning is not done for a florist's pleasure, however, but rather to encourage the fruit that remains on the tree to grow larger. The professionals use their super-trained hands to do this. I use my pruners and agonize over every single cut.

Pest Control

A collar around the tree's trunk can help deter nibbling critters from wreaking havoc on its bark. Once the fruit is on, a disco ball or reflective strips hung from the tree's branches will deter nibbling birds. Bird netting can also be draped over the tree to prevent birds from pecking at the fruit before harvest and is lightweight enough to keep them from getting trapped. I've even heard of some tree guardians that will carefully and painstakingly wrap each fruit in a brown lunch bag to protect it from pesky critters. When you've waited years to get to this point, I get it.

A variety of holistic orchard practices are available to combat common pests. Consult your extension office for more information about pest treatment in your area. If a problem with pests arises, don't leave it alone! Treat it.

Harvest Time

If there was ever a romantic moment on the farm, this is it. You've been waiting for this moment for months . . . perhaps even for years. You've been lovingly watching that fruit grow on the branches each day, starting out as blush pink blossoms and growing into round, crisp apples. It's like magic every single season. How do you know when it's time to harvest?

You reach your giddy little hand up into that tree and you pluck a fruit from the branches, press it to your lips, and take a bite. Hear that? That's the sound of angels singing. Harvest time is always a cause for celebration. Around these parts, it's cherries, plums, and apricots in June, peaches and nectarines in July, pears in August, and apples in September. It's months and months of pure heaven.

Because fruit is extremely perishable, it's important to process what you harvest quickly. Fruit can easily be canned, dehydrated, frozen, and (of course) eaten! Here are two recipes that have become favorites with my family.

Crockpot Applesauce

12 to 16 apples, peeled and cored

1 cinnamon stick

1 tablespoon fresh lemon juice

1. Peel and core the apples.

2. Dice the apples into 1-inch pieces. No need to be precise.

3. Combine the apple, cinnamon, and lemon juice in your crockpot. Cook on low for 6 hours.

4. After the cooking period, simply use a potato masher or wooden spoon to stir the apples and break up any pieces. They'll dissolve into velvety apple goodness. Lawd have mercy!

Pear Butter

3 pounds overly ripe pears

2 teaspoons lemon juice

¼ cup dehydrated whole sugarcane or honey

1 teaspoon vanilla extract or 1 vanilla bean

1 cinnamon stick

5 cloves

5 allspice berries

1. Give the pears a water wash, and using a paring knife, cut out all of the rotten spots and such. Bruises are really no big deal, but we don't want anythin' too grody in here. Be sure to remove the stem and core too.

2. Puree the pears in a food processor or blender.

3. Add the pears to a large bowl and mix in the lemon juice, sugar or honey, vanilla, and spices. Mmm. Then, pour the mixture into a 9 x 13 pan (or two!).

4. Bake the pears in a 300 degree F oven for 2 to 3 hours, stirring every 30 minutes or so. The longer you cook the pears, the thicker the resulting butter will be.

5. Once the pears are thickened to your liking, remove them from the oven. Then *carefully* remove the cinnamon sticks, allspice berries, and cloves.

6. Ladle the pear butter into sterilized glass jars, put on the two-piece lids, and process in a water canner for 10 minutes.

I love pear butter so much I decided to plant a few pear trees on the farm for a surplus each fall!

Potted Fruit Trees

If your climate is anything like mine, it's not ideal for some of the most delicious fruits out there! Lemons, limes, pomegranates, figs, persimmons . . . oh, how I long for you! A dear friend of mine lives in Arizona and likes to brag about the citrus slushies she gets to enjoy all winter long. While we're up here in snow pants eating fruit leather, she's down there enjoying fresh lemonade right from her tree! I've tried to buffer the loss I feel each winter by growing a few fruit trees in pots that I can stick outside in seasons when the weather allows and then bring them back indoors through the colder times of year.

If you want to do this yourself, plan on a 15-gallon container to plant your tree in. The roots still need as much room as you can give them to spread! Obviously, the weight of the container, plus the weight of the tree, plus the weight of the soil can be a serious consideration when you're planning on where to move your tree to and from. Plan on having some buff friends come over to help you bring it indoors in the late fall. Tell them you'll pay them in lemonade.

Great Trees to Grow in Pots

Dwarf citrus

Pomegranates

Figs

Apples

Cherries

Caring For Your Potted Tree

▷ Mix peat moss into the soil to aid in water retention.

▷ Citrus trees thrive with a bit of sand in their soil. Mix some into your potting mix while you're planting it.

▷ Potted plants can dry out quickly when they're outdoors in the warm months. Make sure that you're watering at least once a day on those hot days. Mulching the surface of the soil can help significantly.

▷ Fruit trees do well with consistent fertilization. Be sure to fertilize at least monthly year-round.

▷ Even potted trees benefit from pruning. Clean up the dead or oddly formed branches by pruning them away and encouraging the tree to grow in a strong and stable pattern.

Fruit Wines

Yes, most wine is made from grapes. But since there are already a lot of people doing that (I bet some of them even in your immediate area) and doing it so dang well, why don't we instead invest our efforts in fruit wine? This is a great undertaking for us on our own homestead, as there is always extra fruit at our disposal. Can you imagine sitting in the garden, surrounded by the smell of flowers, and pouring yourself a glass of chilled peach wine? Y'all. This is the stuff that dreams are made of.

All winemaking begins the same way: The fruit is crushed before yeast is added. The yeast feeds off the sugars in the fruit and turns the sugar into alcohol. Fermentation at its very basic level! When Stuart and I began making fruit wine way back when, we made it from store-bought frozen fruit. We mixed a little bit of this with a little bit of that and—bam!—our first peach wine was born. A year later, we trudged down to the dirt cellar, grabbed a bottle from among the cobwebs (seriously, there were a lot of cobwebs in that cellar) and brushed it off. Gently, we uncorked it . . . drizzled a bit into a cheap wine glass . . . and took a whiff of its aromas. Hey, it didn't smell rancid–that was a good sign, right? I let Stuart take the first sip just in case it had turned into poison before I poured myself a small glass. For a first effort, it was pretty dang good! Very enjoyable, as a matter of a fact. We enjoyed the other forty-two bottles that year, taking some of them over to friends' houses to share a bit of our homemade bounty. Looking back, I sure wish we'd written down that recipe. Lest you think it's all unicorns and rainbows, we've made some horrendous wine as well, but we drank it anyway. One simply does not waste thirty bottles of wine.

Even basic fruit winemaking does require a few pieces of specialized equipment that will make the process much easier. But once you've made the initial investment, you'll be prepared to make fruit wine for years to come!

The trick with winemaking is getting the sugars and acids in the resulting wine to balance out in a delicious way.

Basic Supplies

- *Fermentation bucket.* A large, food-grade plastic bucket that will serve as the fermentation vessel for the fruit and the yeast.

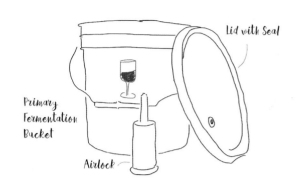

Lid with Seal

Primary Fermentation Bucket

Airlock

- *Large spoon.* For stirring stuff, yo.

- *Glass carboy.* After the wine has fermented, you'll place the resulting liquid in a large, glass carboy where it will settle and age.

- *Airlock.* A gigantic cork of sorts, the airlock allows the fermentation vessel to release carbon dioxide (a byproduct of the fermentation) without letting any oxygen into the vessel. It's quite fun to watch the fermentation bucket "burp" carbon dioxide bubbles.

- *Racking cane/siphon hose.* A super-cool invention, the racking cane is a long glass tube that can fit into the narrow neck of the glass carboy and easily siphon the wine into the bottles.

Glass Carboy

- *Bottles.* Yes, you *can* recycle wine bottles! Or you can buy new ones. Either way, you'll need a stash of them.

- *Corks.* For plugging the bottles.

- *Corker.* For getting those slippery ol' fat corks into the bottles.

- *Hydrometer.* A hydrometer measures the density of liquids. This tool will allow you to easily test the sugar content of your ferment. This allows you to stop the fermentation at the correct amount, depending on the recipe.

Racking Cane and Siphon Hose

- *pH testing strips.* An easy way to test the acid levels in your wine.

- *Acid blend.* If the acid level in the wine is too low, adding a general acid blend (containing citric acid, malic acid, and tartaric acid) can bring it up to the correct level.

> *Sanitizer and brushes.* Hear me now–your equipment must be clean! You're trying to get the wine to ferment but in order to do so, you need to keep bad bacteria and contaminants at bay. Special cleaning solutions and brushes are made specifically for winemaking equipment and are worth the small investment.

The Basics

Now that you have the equipment to make wine, here's how you make it:

Step One: Crush It.

Naturally, start with good fruit! Fruit that is underripe will be acidic and not have the necessary sugars needed for fermentation. Overly ripe fruit will yield a sweet, heavy, and bland wine. Use the best fruit you can find. Pit it. Crush it with a wooden spoon, a tight fist, or your feet. Heck yeah.

Step Two: Mix It.

Because fruit is typically much lower in sugar and acid than wine grapes (read: the product that makes really, really good wine) it's often necessary to add sugar and acid to particular levels, depending on the content your fruit already contains and what the recipe calls for. I recommend organic sugar. At this stage, you'll mix your crushed fruit with the sugar and acid (if needed, depending on your recipe). Then, you'll add enough filtered water to reach the amount of wine you'll be brewing (most homebrews are around five gallons). Grape wines are made entirely from grape juice, but because different fruits contain varying levels of acid, sugar, and flavor, almost all fruit wines are mixed with water to obtain a more desirable balance. At this point, the mixture is left to rest for twenty-four hours in the fermentation bucket.

Step Three: Yeast It.

Sprinkle the wine yeast over the top of the juice. Cover the bucket with a towel and allow it to rest in a safe location, at room temperature, for seven days.

Average Amount of Fruit Needed Per Five-Gallon Batch of Wine

Apricots, 20 pounds

Blueberries, 15 pounds

Blackberries, 16 pounds

Cherries, 5 pounds

Gooseberries, 10 pounds

Peaches, 17 pounds

Plums, 17 pounds

Raspberries, 15 pounds

Step Four: Strain It.

After seven days, strain the pulp from the bucket. This can be done by stretching and tying a large piece of fine-mesh cheesecloth over the top of a second fermentation bucket. Pour the first bucket into the second bucket. This will remove the pulp and strain the liquid. Because, ya know, this step is all about the straining.

Step Five: Second-ferment It.

Yes, we've already fermented it a bit. But now we're going to ferment it some more. So we're going to take that strained liquid from step four and put it into a glass carboy. Here it will sit for about six weeks while it continues to ferment and develop. Plug the top of the carboy with an airlock where the carbon dioxide bubbles will continue to pop up–more rapidly at the beginning and then slowing down as the fermentation process begins to slow.

Step Six: Siphon It.

After the six-week period, a large amount of sediment will have settled to the bottom of the carboy. You'll see a distinct line in the wine. A siphon will allow you to take the wine off the top of the carboy while not disturbing the sediment at the bottom (we don't want that in our wine!). Some winemakers prefer to siphon to a second carboy and allow the wine to settle one more time before finally bottling the wine.

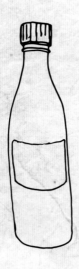

Step Seven: Bottle It.

The siphon will help you do this as well. Or you could break out a good ol' fashioned funnel and do it the hard way. A corker will make corking the bottles easy. (I'll just pretend at this point that I haven't spilt a few bottles in my day from trying to cork the bottles at an awkward angle.)

Step Eight: Age It.

At least six months is the benchmark for fruit wines, but hey, it's your wine. Just know that the wine takes time to develop, so opening it too soon will not be a good reflection of the wine that's actually contained in those bottles! Give your wine a few good months to age (hey, it only gets better, right?) make a new batch during the summer when fruit is at its peak, and you'll be swimming in wine for years to come!

So now, you've made some wine. Go you! While it's beginning to age and collect dust in the cellar, find a ready bottle from a local winery and pour yourself a generous glass to enjoy while you read a bit more about how some dear friends of mine around the country "farm" in their own ways. This certainly isn't a "one size fits all" lifestyle, after all.

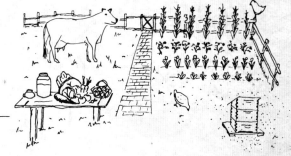

CHAPTER 8

Highlights from Other Homesteaders

Throughout the years, I've had the privilege of meeting and connecting with other homesteaders and real-food enthusiasts from around the country. Though they began as business relationships, many of them have turned into friendships that emphasize the beautiful community of the homesteader movement. As you'll see, there is no longer a one-size-fits-all approach to backyard homesteading. It's different from one farm to the next, and that makes it both beautiful and sustainable. Allow me to introduce you to a few of my friends and their stories in their own words.

Quinn: A Homesteader and Market Gardener from Ohio

We live in a cheap culture. Relationships are shallow and food is inexpensive. Leisure time spent recreating is placed at a premium. We connect with a computer screen between us. We fill our homes with empty calories that fail to nourish, satisfy, and fill the void. We demand everything in an instant.

I believe that many of us are realizing our need for something deeper. Something more fulfilling.

What we are discovering are meaningful connections with our families and the people in our communities, a deep respect and love for the land and God's Creation, and the rich sense of satisfaction that can only come from putting our hand to the proverbial plow and working hard to grow the food that we serve our families. It fills the emptiness of our time and our bellies.

The easiest place to make that connection we yearn for is around the family table over a wholesome meal made even more fulfilling by our labor. Interestingly, it's the way things used to be. The old paths are calling and we are returning to our roots in droves.

We're experiencing the thrill of serving a tomato that goes from the garden to our table in under ten minutes and is sprinkled with fresh herbs we stooped to gather on the way. In the morning we drink deeply of creamy Jersey milk served next to the golden yolks of eggs collected that morning. We can gather around the table and reflect upon the impatient two years it took for that glass of milk since the calf came home . . . Or that the tomato was four months in the making since the seed was sown and the plant tenderly cared for. The freshness and unseen ingredients–blood, sweat, and tears–combine in those moments to create an unfathomable appreciation for all that went into the life that feeds us.

I'm not sure if it matters so much *how* we got started homesteading; what matters is why we continue. Those are the reasons why we head outdoors to weed tomatoes on a ninety-degree day or milk the cow on a ten-degree morning. Many of the reasons to begin were born of romanticism and not reality.

My family started our homestead around ten years ago. It was born of the desire to provide our children with meaningful work that had a return on their labor investment. I was learning more and more about the concerning nature of our modern food system and wanted to check out of it. I wanted my children's bodies to be nourished instead of simply fed. We started with a few laying hens and a small garden. Before long, we figured if we could raise chickens for eggs, why not meat? And then, of course, it's easy to start dreaming of homegrown bacon to go with the eggs at breakfast and washing it down with a cold glass of milk converted into its current state from the grass growing in the backyard. (Because who likes mowing the lawn anyway?)

Today our homestead includes about thirty laying hens, a family milk cow, her yearling heifer, this year's bull calf, and a beef cow to fill our freezer this winter. Seasonally, we have one hundred broiler chickens and a couple of crossbred heritage hogs. This year, we expanded our gardens and are growing intensively on about an acre (including under a three-thousand-square-foot hoop house). Tonight we're going to look at a couple of lambs to bring home. Seems like a lot? Well, we are a growing family of ten!

We grow and preserve much of our food for the year and have been completely independent of outside sources of meat and eggs for many years. I'm not the biggest fan of canning. I'd rather spend my summer days outdoors than in a steamy kitchen, so we try to grow and eat with the seasons. That means that with the exception of a few staples, we go for several months without certain foods. From that experience I've learned that after fasting from a food for so long, when it is time to taste it again during its peak of freshness, the flavor is incomparable to anything that I could have canned and served every week from the pantry. Eating alongside the seasons seems to come naturally with this lifestyle.

The beauty of this life is there are no standards. You don't need forty acres, twenty, ten, or even one acre to get started living the simple life. You don't need anything other than a counter in your home where you live right now to get started. A place where you can chop vegetables to funnel into a mason jar and can them for your shelf; measure some herbs and cover them with alcohol for an herbal tincture; combine a few oils, water, and lye to saponify and make your own soap. A place where you can learn to cook your meals from scratch using a few simple ingredients from a backyard planter or bought fresh from the local farmer's market.

I suspect that you'll find once you've begun to live your life against the grain, you'll discover the nourishment for body and soul you've been craving. And the skills you begin building will feed you for a lifetime.

I believe that Creation and society need us to serve as reminders of how we're here for more than taking what we can get. They need us to serve, give, build, nurture, and steward. We need to be defenders of the patterns of life set forth from the beginning and prevent progress when it depletes resources and destroys life. The true value in homesteading is found when we pass on the skills we are learning to others and especially to our children so that generations to come can know where to go when they need something deeper than our shallow culture gives them. If we've done our job well, there will still be an abundant earth for them to cultivate.

DaNelle: An Urban Farmer in Arizona

Seven years ago we were newbies to the farming world, we didn't know how to grow anything and, truth be told, I was completely afraid of chickens.

I had always wanted to live on a farm, but it wasn't until I was married and had kids of my own that I took the farming dream seriously. At the time we were living in a small condo, and I could think of nothing more glorious than gathering my own eggs and milking my own goats, even though I had yet to even taste goat's milk. I loved the idea of gathering eggs or harvesting veggies, but more importantly, I loved the idea of a slower lifestyle, one that avoided the rush of the everyday and instead focused on appreciating every little thing. We started looking around at lots of land that we could afford, and when an old fixer-upper with an acre covered with mature shade trees became available on the market, we snatched it up. One day, without knowing a thing about animals, we brought home six chickens and one goat, and we've never looked back. The desire to learn how to grow my own food was my driving force, no matter how tough the task.

The first time I milked our goat, I ended up in tears. As I sat wondering how this city girl got in this predicament, I had to laugh at it all. But, I stuck to it and each day I got better and better. And after a few months, I could milk a goat with ease. In our first experience raising chickens, I wasn't sure if they were going to cuddle with me or peck my eyes out. Over time, I learned their behavior was unique and incredibly intuitive. Eggs became such an easy thing to obtain here on the farm, and keeping the hens healthy was even easier as long as I provided an environment that met their instinctual needs. The first time I grew vegetables, I had a few successes, but mostly failures. The best part was that nature was forgiving and I could start fresh again in a few months during the next growing season. And as each season of growing passed, I became better and better at knowing when my plants needed water, knowing when they needed fertilizing, and knowing when to harvest.

I'm still blown away at how providing our own eggs, milk, meat, fruits, and veggies can be so rewarding.

Maybe it doesn't take much to please me, or maybe there's a huge basic human reward in being able to place a seed in the ground and have it turn into a bundle of green beans. Even after all this time, seven years later, I still get a little giddy gathering the eggs each morning. We have ten to fifteen laying hens (those buggers are constantly moving, so it's hard to keep count) that give us anywhere from eight to twelve eggs per day depending on the time of the year. We typically add a few each year because some are lost to weather or predators.

Once a year around August, we purchase thirty meat chicks to raise and butcher. These last us all year. We typically have two milking goats at any given time. If you have a farm, or even a decent-sized backyard, trust me, you need a goat! Our first milking goat is Penny, a Nigerian Dwarf goat. This breed is known for the sweetest-tasting milk. She is a smaller goat and only gives us about one to two quarts a day. Our second milking goat is Luna, a Nubian goat. This breed is known for producing large quantities of milk. She gives us about one to two gallons per day. This is plenty of milk for us! While there are lots of things you can make with goat's milk, we pretty much stick to homemade yogurt and mozzarella, which we like to freeze and have on our homemade pizzas. We breed our goats once a year so they have babies and their milk is "freshened," so throughout the year we usually have a slew of baby goats running around. Goats can have anywhere from one to six kids. Unless we decide to raise one to become a future milking goat, we usually sell the baby goats to local farmers or FFA kids when their mamas have weaned them.

We love keeping sheep on our farm! In the beginning we raised Katahdin sheep, but have recently moved toward an East Friesian mix. East Friesians are known for their milk, and since sheep's milk is my favorite, I plan on milking Eleanor very soon. Once a year, we typically raise a lamb to adulthood, then butcher it for meat. This year we aren't going to, because Eleanor gave birth to a female, and we're going to keep her as an adult. We have recently added Muscovy ducks to the farm. They are great at keeping the bug and mosquito population down as well as providing another great meat source. We purchased one drake and three hens as our breeding stock. We have a five-hundred-square-foot garden, and we will also grow vine fruits and veggies in another part of our yard. In the spring we produce snap peas, carrots, parsnips, lettuce, spinach, chard, kale, onions, garlic, celery, tomatoes, peppers, cabbage, broccoli, sunflowers, cilantro, and parsley. In the summer we grow green beans, potatoes, tomatoes, peppers, cucumbers, squashes, watermelons, cantaloupes, and basil. In the fall and winter we're all about corn, carrots, lettuce, spinach, chard, kale, squashes, pumpkins, eggplants, cauliflower, cabbages, broccoli, sweet potatoes, watermelons, cantaloupes, dill, and mint. We have more than fifteen different fruit and nut trees or plants. We have grapefruit, blueberries, strawberries, mulberries, loquat, peaches, blackberries, grapes, mangoes, cherries, figs, bananas, almonds, avocados, pecans, and oranges.

In a nutshell, backyard farming teaches you that you can only do the best you can, and there's no point in dwelling on the past, only learning from your mistakes so you can be better in the future. Growing your own food changes you, in more ways than one. You learn how to have patience, to develop a thirst for knowledge, and to accept that you have no control over the success of your farm, only to work hard and hope for the best. Above all, you learn to laugh at yourself and appreciate homegrown food in a way you never did before.

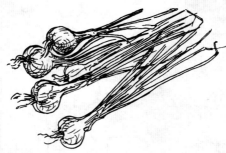

Angela: A Small-Scale Homesteader and Food Enthusiast from Illinois

I homestead because I can't imagine life any other way. It amazes me that it hasn't always been like this for me—for us a family. There was once a time when we lived in town, on the perfect corner lot, and I scoffed at the idea of growing anything but my beloved perennials. But hard times hit and I realized I could help our growing family save a little money by growing vegetables.

Soon after, we bought some chickens (which were illegal in our town) and all the while this change was happening in our hearts. Each day, we awoke with the need in our souls for more space, more produce, more animals, and more room for our children to run. The neighbors finally reported us for our chickens and that same day we put up a homemade for sale sign. We were stunned at who we had become. It was time to move to a farm.

Our transition was unique because we sold our house and moved to a rental farm. We decided to "practice" homesteading while we put away money for our dream farm. We found a sad, little slightly charming three-acre property that hadn't been used as a farm in years, and worked out negotiations with the landlord to turn it into something fabulous. He allowed us to do minor improvements to the farmhouse and to build our own chicken coop and dig a huge French-style potager or kitchen garden. All before we even moved in. We named our homestead Half-Way Farm or *la Ferme a Mi-Chemin*, because it's halfway down our beautiful rolling road on the Wisconsin/Illinois border and because it's halfway to our dreams of owning our own.

Each passing season has taught us something as we prepare to move next year. Having gardened only on a corner lot in town, I had to learn a few things and sometimes the lessons were rather frustrating. The first year at the farm I had a horrible crop of tomatoes, cucumbers, and peppers. I found myself completely baffled that these vegetables that I had grown for years with much success were stunted and wilting. After much

research, I discovered that the black walnut trees that cover this property release a chemical into the soil, making it impossible for certain things to grow. We build raised beds for those particular vegetables and had uncontaminated soil brought in. Lesson learned, and now we know not to buy a property with black walnut trees!

We've also raised plenty of meat birds. Our plan was to raise two varieties back to back and compare the results. We began with Red Rangers and then moved on to the more popular Cornish Cross. We fell in love with the Rangers, finding them more like an actual bird, less messy, and fabulous free rangers. In an effort to become even more sustainable, our next experiment is to utilize our rooster and simply hatch and raise the heritage breeds that we use for eggs. And if you had told me ten years ago that I would be butchering my own chickens, I never would have believed you. At the time I could barely even touch chicken meat without feeling creepy!

What I've strived to create is a romantic homestead with a distinctly French flair. The layout of the gardens, charming pea-gravel paths, and garden architecture remind me of the delightful farms scattered all over the French countryside. It's peaceful here and the air smells of farm life and herbs.

Each year we find ourselves diving into a new project, taking on a new animal, or wanting to learn a new skill. I got it in my head one winter while being stuck in bed with the flu that I wanted to raise goats, Nigerian Dwarf goats specifically. What mama wants, mama gets! We began our herd with Coco and Valentino, and have since enjoyed our first kidding and are expecting more little ones here soon. I spent years of my life considering myself "not an animal person," but now my mornings are filled with the bleating sounds of six goats, countless chickens, and a very proud, very flamboyant rooster.

There are things we can't do on this homestead because we are renters. I can't have the beautiful cow I've dreamed of, and I'm trying to keep our little goat herd down in numbers, knowing that moving day is just around the corner. I did decide to add two things, however, to our farm recently without asking permission. We are now proud owners of two beehives and two Berkshire pigs that are fattening for the freezer this fall. There's something wonderfully contagious about keeping bees. Once the desire strikes it's all downhill from there. I will say it's been the most daunting endeavor as it's an attempt to partially tame and understand creatures that are 100 percent wild and mysterious. The first year our hives died, but I have high hopes for our next attempt. Even if I end up with one jar of honey in my pantry this fall I will beam with pride.

Having six children, the need for food is always there. What better way to bring a family together than to learn the art of hog butchering? I love to just jump right in and learn on the job, so I was thrilled when I found my Berkshire feeder pigs on Craigslist this spring. I was very pregnant at the time, and so loading my kids in the truck and driving out for those pigs on a whim was a fantastic way to make a memory and to learn about pigs, a trial by fire.

This rental farm experiment in homesteading has confirmed that we are country people at heart. We'll never go back to life in town and we are counting down the days until we buy . . . the big one.

Jenny: A Real-Food Cook and Supporter of Small Farms

I am not a farmer. In truth, I can barely coax a tomato from my garden.

I linger late in bed in the mornings whenever I can. I travel on a whim. I have no patience for tending a laboring ewe beneath a cold and star-pierced sky into the impossibly dark hours of early morning. And the idea of mucking out a chicken coop leaves me queasy. I once managed to pluck our Thanksgiving turkey, but my dinner guests had to finish the job, pulling pinfeathers out of that golden, roasted skin, and delicately sliding them to the edge of the plate while we all graciously pretended not to notice.

Let's just leave it at this: Farm work isn't for me.

Yet, there's a longing I have for farms and that idyllic, pastoral life. Even fully conscious of my woefully poor aptitude for seemingly all agricultural undertakings, I find myself imagining a time when I might have goats, or raise a few ducks for eggs, and then reality sets in.

Instead, I choose to support the men, women, and families who run their farms, ranches, and dairies tirelessly, skillfully, and with an unyielding stewardship that few other professions know. There's a dance that happens between small-scale farmers and their well-wishers. That is, the economic viability of homesteading depends, at least in some part, on the farm-to-consumer relationship. One depends upon the other.

You don't have to own a farm or homestead to support those that do.

In urban and suburban communities, that support takes the form of farmers' markets and community-supported agriculture shares (CSAs). A CSA allows individuals and families to purchase a share of a farm's harvest. Shareholders typically pay in advance before the harvest season begins so that cash-poor farmers can purchase seeds and equipment for the coming year. The farmers, in turn, provide their CSA members with a portion of the harvest, one that's abundant in years of plenty and a bit leaner in years affected by drought, unpredictable weather patterns, or other hardship. It's a commitment to care for and to be cared for, which is the root of community.

Farmers' markets provide an opportunity to connect directly with growers and food producers. Urban farmers' markets can prove particularly lucrative for farmers who typically travel from rural farming communities where there's little market for their goods, since everyone seems to grow their own or knows someone who does.

For shoppers eager to connect one-to-one with growers and forge that farm-to-consumer relationship, farmers' markets present a fantastic opportunity, but it's wise to ask questions not only of the growers and vendors, but also of the market itself. In less rigorously managed markets, growers and producers are mixed with vendors who resell their wares or have no firm regulation that limits what can be sold or where it must be produced.

For shoppers in search of authenticity, such markets can leave them wanting to swear off farmers' markets altogether. After all, if there's no guarantee that the person selling you those cherries really did grow them or that they even came from a nearby orchard, what's the point? The solution is to look for producer-only markets, or, in states that offer certification, look for certified farmers' markets, as these markets disallow the practice of brokering or purchasing fruits and vegetables wholesale and selling them to the consumer. They ensure that the person selling you your emerald-green bunches of collards and blushing pink apricots really did grow them. Also, communicate directly with growers when you can, asking them which methods they use to grow, whether they spray their crops, and how their animals are cared for. Ask if their farms are available to visit, and make a trip to the countryside to check them out on a day they host farm tours.

If you have a day or a weekend to spend in the countryside, you might stumble upon a u-pick farm or garden, where you can pluck sun-ripened peaches in summertime or crisp red apples in autumn straight from the trees. Children particularly enjoy the opportunity to run and play outside beneath the awning of fruit trees, gathering what they can into baskets. That direct connection with the grower also allows you the opportunity to ask questions about how the food is grown, and what kinds of inputs (fertilizers, chemical pest controls), if any, they use.

In small towns and rural settings, if you skip the highways to slow down and take the scenic routes, you'll come across small farm stands at the ends of driveways selling produce, and sometimes eggs in coolers or mini-fridges. These stands almost always operate unmanned and on the honor system. You take what you need and you drop your cash into the cash box.

In urban communities, you can often find community gardens. These communal plots of land offer those who don't have the space or resources an opportunity to get their hands into the dirt to sow seeds and reap the surprisingly abundant harvest that can come from a 10- by 10-foot plot. Moreover, they give you the opportunity to rebuild that beautiful sense of community, which includes sharing tips, working together, and bartering.

While resources like CSAs, farm stands, farmers' markets and community gardens are plentiful and growing, in some areas those resources have yet to find their place. This gives you the unique opportunity to shape the food community that you want to see.

When my husband and I wanted a farmers' market where there was none, we built one. When the CSA we wished to support couldn't service our town due to lack of interest, we found members. There's always the opportunity to build community where there was none before, and, in doing so, you build the framework to support the work of farmers in a very real way.

You don't have to be a farmer to support their work.

RESOURCES

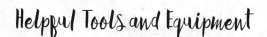

Helpful Tools and Equipment

Farm Tools

I'm a minimalist when it comes to farm tools. You'll likely see me out in the garden on my hands and knees, digging rows for my beets with the edge of a sharp rock. Seriously, I've done it. Part of this comes from my frugal nature and the "adapt and overcome" mentality that I have as a farmer. The other part of it comes from having four children and never wanting to load them all up in the car, take a trip into town, and enter a store with my circus of humans for a new shovel. Our first farm was mountainous and steep, in no way lending itself to a tractor of any sorts. Plus, we were poor! It was the perfect storm for learning how to make-do or do without.

Lest you feel like you need to have a big ol' barn filled to the brim with equipment to make this whole farming thing happen, fuggedaboutit. You'll be just fine with your own two hands, a hammer, and a dream. Alright, fine, you may need a bit more than that. But I've paired the list down to the "essentials" as I see them at this moment. I'm sure your list will look different, and that's just fine. Take note of what you use, what you don't, what works, and what doesn't. Invest in good equipment–buy the very best that you can afford. Cheap equipment breaks, usually right in the middle of a project, and will only cause frustration and extra trips into town.

For the Animals

▷ *Thermometer.* One of the easiest ways to tell if you need to call in the vet is to take the animal's temperature. Any idea where you stick that thermometer? Yep, that's right. Right in its pasty butt. Don't worry, you're surely not going to use the same one that you use on your kids. Have a separate one in the barn for such occasions.

▷ *Apple cider vinegar.* Apple cider vinegar is a great little electrolyte boost for your animals and can serve their health well added to their water each day. A few tablespoons will do it. Having a big ol' stash in your barn is better than money in the bank.

▷ *Molasses.* Sometimes when animals go down with illness, they need a "drench" to help get them back up. Molasses is a great source of sugar and minerals for animals and can give them a huge boost whether it's poured over their oats or watered down. We like to feed our new mamas a bit of molasses in their water to keep them perky after all the hard work of birth!

A large stash of hay will make it possible to keep your animals fed through the winter when grass no longer grows.

▷ *Hay and feed.* Depending on the size of your homestead, the size of your feed stash will vary. We feed hay here a significant portion of the year, which requires us to stock up on hay each summer when the fresh bales are coming out of the field. On top of that, you can often get a better price on grain if you buy in bulk, so it may be worth it for you to stock up on your laying feed or pig oats. Animals don't take too kindly to not being fed on time, so make sure you've got lots on hand.

▷ *Water troughs.* Animals drink a lot of water and need to have it on hand at all times. Invest in strong and big water troughs for them! We like the metal ones best. When you're picking the size for your space, be sure to think about the animals that will be drinking from it. For example, a tall trough won't work for lambs. Also, chickens will most surely fall into any water that is deeper than their bellies. Spend a bit of time thinking through what will work best.

▷ *Bedding.* Dry and clean bedding, dry and clean bedding, dry and clean bedding! The quickest way to get sick animals in the winter is to keep them on cold, wet bedding. Did I mention it should be dry and clean? There are many bedding options available. Straw is my favorite. Have a stash on hand so when you're cleaning out pens, you've got fresh bedding to put back in. Animals thoroughly appreciate deep bedding.

▷ *Bone saw.* As you begin to butcher animals on your farm, having a high-quality bone saw is extremely valuable. I've used some cheap ones in my day and they're simply horrendous to use. Invest in a good one and it will serve you well for decades to come.

▷ *Pig scrapers.* Often available at antique stores, the pig scraper . . . well, it scrapes the pig. It's nothing more than a bell-shaped cup on the end of a wooden handle, but it works very well! And come slaughter day, you'll certainly be glad you have a few in your stash.

▷ *Propane burner.* The easiest way to heat large amounts of water for slaughter day is with a large, powerful propane burner. It's easy enough to stick a metal 55-gallon drum over the top for slaughter day.

▷ *Chicken plucker.* If you're going to be raising meat chickens, having a chicken plucker is awesome. Not necessary, but awesome. These are easy enough to build yourself or are often available from friends to rent or borrow for the weekend. There's also a chicken plucker drill attachment that can be pretty dang efficient at getting those pesky feathers off. You know what else makes great chicken pluckers? Small children that need a task to keep them busy.

▷ *Kill cones.* These make chicken slaughtering very easy. The chicken is hung upside down in the cone so its head sticks out the hole in the bottom, where the throat can easily be slit and the bird can bleed out without flopping all over the farm.

In the Kitchen

- ▷ *Pressure canner.* A valuable tool for both preserving and cooking food. I always err on the size of the largest size I can get!

- ▷ *Water canner.* Another valuable tool for preserving, the water canner can also double as a dunking pot when you're scalding your chickens!

- ▷ *Dehydrator.* Try to find the biggest dehydrator that you can! The large ones will average around 10 square trays and will serve you well. This is valuable for dehydrating fruit, herbs, even jerky.

- ▷ *Large stockpot.* How are we going to boil our pig's head for headcheese if we don't have a large stockpot? Seriously though, my stockpot never leaves my stove. It's always full of chicken stock, large batches of baked beans, stews, or preserves.

- ▷ *Good set of wooden spoons.* Forget fancy gadgetry–a solid, strong set of wooden spoons will serve you well. Plus you can use them to chase after chickens that are scratching up your garden.

- ▷ *Set of sharp knives.* I keep a small clutch of knives in the kitchen. Many of them I've had for a decade. They're my right hand for all kitchen tasks. A few good chef's knives and a small paring knife are all you really need. All of mine are wet-stoned sharpened each week.

My knives are the backbone of my kitchen and are wet-stone sharpened weekly.

- ▷ *Cutting boards.* I prefer thick, wood cutting boards. They are treated with a beeswax treatment every so often to prevent them from drying out and are disinfected with salt and lemon juice.

▷ *Butcher paper.* Butcher paper is the perfect wrap for protecting cuts of meat in the freezer. Have it on hand and you'll find plenty of ways to use it.

In the Barnyard

▷ *Wheelbarrow and/or wagon.* There's this magic thing that happens on a farm—nothing is ever where you need it. A strong wheelbarrow or wagon will make shifting, moving, and hauling things much easier. If you've got more acreage, a small golf cart or four-wheeler will also serve this purpose well.

▷ *Muck rack.* Call me crazy, but I love mucking stalls. It's fulfilling to whisk away the gross waste and replace it with fresh bedding. Sigh. That makes for a happy farmgirl. A muck rack is where it's at.

▷ *Buckets.* If you had no less than 63,183 buckets on your farm, you'd still find ways to use them all. What is it about buckets? Hauling feed, scooping compost, collecting waste, gathering weeds, feeding the dog, you name it. Keep a variety of sizes and materials.

A small rake—perfect for scraping up weeds or chasing away rogue roosters.

▷ *Shovels.* Round ones, square ones, triangle ones. Have 'em all and have 'em ready!

▷ *Rakes.* Great for raking gravel and chicken poop and grass clippings and trimmed tree branches and everything in between. Also great for protecting yourself from crazy roosters.

▷ *Hoe.* A gardener's best friend, the hoe will save your back in the garden by helping you dig rows and whack up weeds.

▷ *Tiller.* I lived without a tiller for many years, but when I finally caved, I wish I would've done it years ago. This makes it super easy to work all that delicious compost into your garden bed and even weed between rows of produce during the summer.

▷ *Drill.* One of the first grown-up tools my husband and I bought when we had a little extra money was a high-quality electric drill. We have run that poor drill into the ground and yet it still gives. My only recommendation: have backup batteries because you'll chew through the charge easily on farm projects!

▷ *A sharp knife.* Always keep a knife in your pocket. Always! Because as soon as you forget to, you'll realize how often you need it!

▷ *An ax.* For chopping wood, yo. And sometimes for putting an injured chicken out of its misery.

▷ *Staple gun.* A staple gun makes putting up chicken wire and protective netting a breeze!

▷ *Variety of nails and screws.* Just keep a huge stash on hand because you'll never know what type or how many you'll need.

▷ *Rope.* No matter how much you have, you'll always need more.

▷ *Tractor.* On the small homestead, it's not necessary, but it's still helpful for just about everything. Do some research or ask an experienced farmer what size would work best for your farm's situation based on weather, topography, and needs.

▷ *Extension cords.* For, ya know, extending stuff like heat lamps.

▷ *Ladder.* Ya know, for reaching things. Like when your turkey has decided to roost on top of the shop and you need to get it down.

▷ *Pallets.* Sometimes you'll need to be able to throw up a quick wall, shelter, or pen. Pallets can easily be tied together with twine or bungee cords in a variety of shapes and sizes. They're also helpful for lifting hay and feed up off the ground where it can get soggy and wet.

▷ *Hose and sprinklers.* For irrigating pastures, gardens, and fruit trees. Also for setting up in the summertime so kids can run through the sprinklers naked.

▷ *General toolbox with wrenches, sockets, hammer, etc.* As a farmer, you're also a carpenter. Keep a large toolbox stashed with all-purpose tools to help you in your new profession. Wouldn't it be nice if a toolbox was stocked with exactly what you needed and was kept organized? Come on! That's like a dream.

▷ *Post hole diggers or auger.* If you've got any sort of livestock, you'll most certainly be sinking fence posts. This means digging fence post holes. And this means you'll need some post hole diggers or an auger to get the job done. A gasoline-powered auger will obviously get the job done quicker and easier than the manual post hole diggers, but also may be more difficult to use in rocky conditions.

I wear many hats on the farm. Post hole digger is not one of them.

▷ *Chop saw.* One of the most versatile power tools, the chop saw will serve you well for many projects (such as when you need to cut boards for your new pig shelter or take the excess off the top of fence posts).

▷ *Generator.* Depending on your power situation, a generator may be an absolute must on your homestead. If you've got a few chest freezers full of meat and lose power, a generator could prevent you from huge losses.

▷ *Fence posts.* Keeping a stash of fence posts on hand can be very helpful for when a new project inspiration strikes! Whether it is building a new run for the chickens, setting up a new grazing pasture for the cow, or even fencing in your garden beds to protect them from the cats, you're going to always find random ways to use up fence posts. Having a stash on hand is extremely handy.

▷ *Pickup truck.* It's time to admit to yourself that you are a farmer and just get that ol' pickup already! When you're hauling compost from a local restaurant or picking up straw hay from the feed store, you're going to really wish you had a truck to put it in. Trust me on this one—I've hauled plenty of animals in the back of my old minivan. And had the stains to prove it.

It's time to wear out your own pair of gloves!

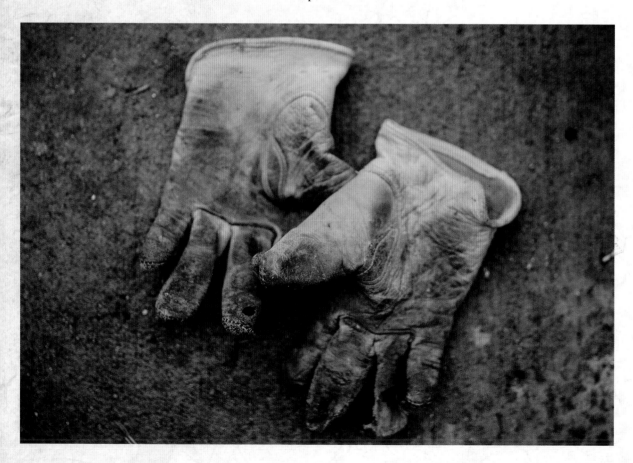

AFTERWORD

Thanks for Coming to the Farm

As I write these words, I gaze out my bedroom window across the pasture. We're at the height of summer and the sprinklers are almost constantly running with their steady and rhythmic paces. The beautiful smell of wet grass is drifting in through the cracked window. If I squint, I can see the tom turkey perched in the chicken coop window where he roosts at dusk to protect his ladies from the owl that continually pays the farm a visit in the wee hours of

the morning. I've got our newest baby to my left, little Juliette, who's wrapped up in a crocheted blanket. To my right sits a home roasted cup of coffee. I can still smell the faintest hint of supper in the air and can hear the voices of my two oldest children bickering quietly about what sort of habitat they should build for all the hornworms they pulled off the tomatoes earlier this afternoon. There are no less than four loads of laundry perched atop my dresser that need to folded and the garden weeds I can see from the window are giving my beets and eggplants a run for their money. But we're here.

We're here, each morning, throwing slop to the pigs and moving the growing bunnies around the yard for grazing. We're here, building fences and new lambing quarters. We're here with sore backs from watering, weeding, and harvesting the garden. We're here, planting patches of comfrey and training the wine grapes up the trellis. We're here, nestled around our nine foot dining room table, breaking bread three times a day. We're here, raising animals only to take their life that ours might continue. We're here, milking cows and making butter. We're here, fighting to take a stance on food production and savoring the blessings that come with farm life.

Imagine there are no confines to your life . . . no fence lines . . . no expectations. Let your mind run with me for a moment, won't you? Where would it take you if you stripped away the assumption that the way your life is now is the way it always has to be?

Years ago, when I asked myself that very question, I was working at an insurance office. A coworker of mine would sit in the back room during every break and lunch hour, knitting a variety of different garments– including little booties for my first child. I watched her as she slaved over a single garment. Because it brought her such joy, the pain of the work didn't seem to bother her. I spent lots of time talking with her, as I struggled to knit my first (extremely ugly) dishcloth. We chatted about gardens and preserving and knitting. She shared an affinity for that taste of hard work and the feeling of satisfaction and reward it brings. We often spoke of what it would be like to live on a small farm and spend our days in the kitchen, kneading bread, and out in the gardens, weeding the rows. Labor, yes, but good labor.

Those days of dreaming with her over my ugly knitting projects has never left me. Rather, just like a seed, they sat in my soul and continued to faithfully grow. It all started small–learning to can, making my first loaf of bread, and starting a compost pile. So my advice to you is no matter your current circumstance, plant and nourish those seeds, baby! Here are some easy ways to get started:

Start cooking. Most of homesteading is done for the love of good food. The very best way to savor the hard work involved in each bite is to start learning to cook that food well. Even homegrown food will taste bad if not prepared properly, so put thought, consideration, and work into learning how to be a rock star in the kitchen. It doesn't have to be complicated, nor should it be. Simply throwing away your processed food and prepackaged

suppers will be a huge step in the right direction. But the cooking? Ah, the cooking is where the joy is found!

Start enjoying fellowship. This means inviting others over for supper and sharing the joys of your harvest! This could be cooking up a big pork feast as a thank you to those who have spent time helping you butcher pigs, or even just inviting friends up for pie and espresso after having spent the morning gleaning pears. Sharing in the joys and stories that homegrown (and often hard grown) food bring along with them is one of the very best ways to appreciate it thoroughly. There's nothing I love more than welcoming another family into our home and breaking bread—even if it's literally just bread, cheese, and wine for supper. Fellowship makes it all taste sweeter.

Build your community. There is a huge community that naturally congregates around agriculture. Guess what? It's time for you to be a part of it. Head down to your local Farmer's Market and spend time talking with the local cheesemonger. Get to know any growers on your road who have produce they'd like to trade or other items to barter. Support them in their efforts and share some of yours with them. This is how a strong and healthy community is built!

Sip and savor. I've learned to do this over the past few years . . . mostly because it's such madness with the little ones that I've got to just take a step back and really soak it all in. The world can be in shambles at my feet, chaos swirling about, the walls crumbling in, and yet I can still sip and savor a glass of local wine while munching on cherry tomatoes, bread, and olives. On a farm, there will always be chaos. Your boots will be muddy and your hands will often be stained with dirt. And yet

if we don't take the time to appreciate our efforts, we'll miss the beauty in it.

Put on your apron. I often joke that my apron is my Superman cape. It makes work easier, gross things less disgusting, and all things possible. The bottom can be held up as a makeshift basket for produce or eggs. The corners of the skirt are good for wiping away a toddler's tears or . . . fine, I'll admit it . . . boogers. You'll always have a place to dry your hands while you're cooking and a few extra pockets to shove seed packets, knives, or wine corks in. An apron says, "I'm here. And I'm ready to make you something delicious. Welcome."

Welcome the challenge. You're going to have days that get you down. When a ewe won't accept her lambs . . . when a chicken gets picked off by an owl . . . when the dog scratches up all your broccoli starts . . . when the bread doesn't rise. It will all come, but rest assured, it will all pass. We once lost a lamb to starvation and as I sat on the ground and held his little, white body I wept and rocked back and forth. Looking up at the skies I shouted, "Who would choose a lifestyle that can cause so many pain? I hate this! It hurts!" But those days get lost in our history and even that horrendous pain somehow makes the good days and small victories so much sweeter than you could ever imagine. Much like labor, we must push through the pain to enjoy the blessing of our hard work.

We have the unique opportunity in this day and age to be a part of something great. We don't *have* to grow our own food, but we *have the freedom* to provide that for ourselves. Maybe for you, that's a barrel of tomatoes on your front porch or perhaps a coop with a small handful of layers in your neighborhood backyard. Yet for others, it could mean making the drastic change to country living and welcoming livestock with open arms and hungry bellies. It doesn't matter *how* you find yourself on the farm. It only matters that you're here, seeking wholesome food and developing an even greater appreciation for it.

Thanks for coming to the farm, my friends. It's a beautiful, and delicious, place to be.

INDEX

Music garlic, 104
mustard greens, seeds, 70

nails, 298
neem oil, 49, 51, 52, 155
neighboring, 180–83
New Zealand Whites rabbit, 206
nipple, pig, 217
northern fowl mites, 153

Olive-Oil Drops recipe, 92
once-a-day milking, 172–73
onions: braiding, 107; cold storing 108; freezing, 91; peppers and, 48
open bin composting, 6
orchard, harvest, 254–70
oregano, 33
organic, feed bag items, 124
organic seeds, 19
oven drying, herb, 144

pallets, 299
parsley, 33–34
parsnips, cold storing, 108
pasty butt, chick's, 148
pâté recipe, 224
pea, seeds, 69
Pear Butter recipe, 263
pears, cold storing, 108
pepper, seeds, 69
Peppermint Tea, 142
peppers: freezing, 91; onions and, 48
pest control, 46; birds and, 48; collar, plant, for, 50, 51; companion planting for, 48; crop rotation for, 50; diatomaceous earth as, 47–48; fruit trees, 264; hand picking for, 49–50, 51; insecticidal soap for, 50, 51, 52; natural methods, 47–50; neem oil for, 49, 51, 52; poultry, 154, 153–55; petroleum jelly, 155; row covers for, 48; summer garden and, 47–52. *See also* individual plants.
pesticide: all-natural, 49; seeds and, 19

Pesto recipe, 94
pests, poultry, 154, 153–55
pH food factor: pressure canning and, 83; water canning and, 77
pH, soil test, 40
pickled asparagus recipe, 79
pickup truck, 300
pigs, 211, 212; bacon and, 211, 212; breeding, 211–12, 218–19; butchering, 219–24; calendar, 218; curing meat, 225; deworming sow, 219; evisceration of, 224; farrowing, 218–19; feathers, 222; feeder, 211; feeding, 215–17; feet, 222, 223; harvesting, 215–17, 219; head, 222; heart, 222; housing, 215; kidney, 222; liver, 222; meat, 211–12; nipple, 217; piglets, 211, 219; preserving meat, 225; scrapers, 294; sows, 218–19; storing meat, 225; as tillers, 59; types, 211, 212, 213–14; water for, 217; weed control and, 45
plant collars, 50
plant, requirements of, 16
planting: interval, 17; maximizing productivity, 16–18; requirements for, 16; schedule, 17
pollination: cross-, 254; of plants, 20; self-, 254
post hole digger, 299
potager, xvii, 4
potato beetle, pest control, 51
potatoes: cold storing, 108; pest control, 51; recipe, 92
poultry pests, 154; dust baths and, 154; lice, 153; northern fowl mites 153; preventing, 154–55; red roost mites, 153; scaly mites, 153; signs of, 153
pressure canning, 83–85, 295; basics, 84; method, 84, 85; recipes, 87–89
produce: freshest, 2–3; livestock and, 3; organic, 3
productivity, maximizing planting, 16–18
protein, homegrown, for chickens, 125
pruning, fruit trees, 258, 264
pullet, 132

Quinn, homesteader, 274–77

ABOUT THE AUTHOR

Shaye Elliott is the founder of the blog *The Elliott Homestead*, which she began in 2010 when her farm was but a dream. She and her husband, Stuart, are now developing their own little farm in the Pacific Northwest with their quiver of children, Georgia, Owen, William, and Juliette. Shaye spends her days writing, gardening, child and chicken wrangling, cow milking, pig wrestling, wine sipping, and dreaming. She has authored two cookbooks, *From Scratch* and *Family Table*, from her farm kitchen.